Benign and Pathological Chromosomal Imbalances

Benign and Pathological Chromosomal Imbalances

Microscopic and Submicroscopic Copy Number Variations (CNVs) in Genetics and Counseling

THOMAS LIEHR

Amsterdam • Boston • Heidelberg • London
New York • Oxford • Paris • San Diego
San Francisco • Singapore • Sydney • Tokyo

Academic Press is an imprint of Elsevier

Academic Press is an imprint of Elsevier
525 B Street, Suite 1900, San Diego, CA 92101-4495, USA
32 Jamestown Road, London NW1 7BY, UK
225 Wyman Street, Waltham, MA 02451, USA

Notice

British Library Cataloguing-in-Publication Data
A catalogue record for this book is available from the British Library

Library of Congress Cataloging-in-Publication Data
A catalog record for this book is available from the Library of Congress

ISBN: 978-0-12-404631-3

For information on all Academic Press publications
visit our website at elsevierdirect.com

Typeset by TNQ Books and Journals Pvt. Ltd
www.tnq.co.in

Cover Figure: Interphase and metaphase after C-banding in background of the title
shows an example of a standard cytogenetic approach for the characterization of
microscopically visible CNVs. In the foreground there is a heteromorphic chromosome
22 which is shown in 'black-and-white banding pattern' (inverted DAPI-staining; left)
and after 3-color-FISH (right); the latter revealed that the massively enlarged short arm
was due to a partial triplication

DISCLAIMER

The clinical details given for specific chromosomal imbalances and rearrangements, including such regions causing, according to current knowledge, no harm, represent the presently available data. It can be used for interpretation of cytogenetic findings; however, exceptions from the findings are always to be expected and some are already described in this book. Use this information carefully; the author does not take any responsibility for (mis)interpretation of the data provided in this book.

CONTENTS

BIOGRAPHY

Dr Thomas Liehr, University Hospital Jena, Germany.

Dr Liehr's laboratory is one of the leading molecular cytogenetic laboratories in the world and has developed numerous specific multicolor-fluorescence in situ hybridization based probe sets. During the last 20 years, Dr Liehr has studied and co-issued cytogenetic reports on over 5000 cases with cytogenetic aberrations. He is an Editor of *Molecular Cytogenetics* and on the Editorial board of *Journal of Histochemistry and Cytochemistry* and seven other journals. Dr Liehr has published ~400 peer reviewed articles on inherited and acquired marker and derivative chromosomes, karyotype evolution, epigenetics including uniparental disomy, interphase and chromosome architecture as well as probe set-development.

aCGH	array-based comparative genomic hybridization
AS	Angelman syndrome
BAC	bacterial artificial chromosome
bp	basepair
BFB	breakage-fusion-bridge
BWS	Beckwith-Wiedemann syndrome
CBG	C-banding; that is, centromeric regions of chromosomes are stained
cen	centromere
CER	centromeric repeat-satellite DNA
CG-CNV	cytogenetically visible copy number variant
CGH	comparative genomic hybridization
CMT1A	Charcot–Marie–Tooth disease Type 1A
CNV	copy number variant
Cytoband	GTG-light or GTG-dark band at a certain chromosomal resolution
DAPI/DA	4,6-diamino-2phenyl-indole/distamycin A
del	deletion
der	derivative chromosome
dic	dicentric
dup	duplication
DNA	deoxyribonucleic acid
EV	euchromatic variant
FISH	fluorescence in situ hybridization
FoSTeS	fork stalling and template switching
GRCh37	Genome Reference Consortium Human genome build 37
GTG	G-bands by Trypsin using Giemsa
h	heterochromatin
hg	human genome
HIV	human immunodeficiency virus
HNPP	hereditary neuropathy with liability to pressure palsies
HOR	higher order repeat
indel	insertion/deletion variants
inv	inversion
inv dup	inverted duplicated shaped sSMC
ISCN	international system for human cytogenetic nomenclature
kb	kilobasepair
LCR	low copy repeat
LINE	long interspersed nuclear element
LTR	long terminal repeat

Mb	megabasepair
MDD	microdeletion and microduplication (syndrome)
MG-CNV	submicroscopic copy number variant detectable by molecular genetics
min	centric minute shaped sSMC
MMBIR	microhomology-mediated break-induced replication
mos	mosaic
NAHR	nonallelic homologous recombination
NCBI	National Center for Biotechnology Information
NGS	next generation sequencing
NHEJ	nonhomologous end joining
NOR	nucleolus organizing region
OMIM	online Mendelian inheritance of man (database)
p-arm	short chromosome arm
pter	terminal end of a short chromosome arm
PHP	pseudohypoparathyroidism
PRINS	primed in situ hybridization
PWS	Prader Willi syndrome
q-arm	long chromosome arm
qter	terminal end of a long chromosome arm
r	ring chromosome
REP	repetitive
RNA	ribonucleic acid
rRNA	ribosomal ribonucleic acid
s	satellite
SINE	short interspersed nuclear element
ss	double satellite
SNP	single nucleotide polymorphisms
SRS	Silver Russel syndrome
sSMC	small supernumerary marker chromosome
stk	satellite stalk
TND	transient neonatal diabetes
UBCA	unbalanced chromosomal abnormality
UPD	uniparental disomy
UCSC	University of California, Santa Cruz
VNTR	variable number tandem repeat
XIST	X inactivation-specific transcript

Some books have to wait for the right person to write them and this volume is one of them. Dr Thomas Liehr is uniquely placed to pull together this comprehensive survey of variation in the human genome, which includes many exceptions to the rule that microscopically visible alterations invariably have phenotypic consequences. Most chromosome abnormalities do have clinical effects but, as the resolution of chromosome analysis has increased, so too have the number of instances in which cytogenetically visible changes are benign.

Dr Liehr and his colleagues in the Institute of Human Genetics have made the Jena University Hospital, Friedrich Schiller University at Jena an international center for molecular cytogenetics. In particular, they have built on the pioneering work done on the microdissection of chromosomal segments; these can be converted into multicolor banding probes with which to investigate some of the most complex and variable regions of the human genome. Dr Liehr is a recognized authority on small supernumerary marker chromosomes, the interpretation of which has presented technical and clinical difficulties for decades. His web site (http://www.uniklinikum-jena.de/fish/sSMC.html) provides a vital resource on the nature and variable clinical consequences of marker chromosomes. The web site is particularly useful at prenatal diagnosis when limited time makes accurate and accessible information invaluable.

This work has been followed by an interest in the phenomenon of uniparental disomy (UPD), which can occasionally accompany a marker chromosome, and the creation of another web site dealing with UPD (http://www.fish.uniklinikum-jena.de/UPD.html). The marker and UPD web sites receive hundreds of visits per month and the appreciative comments from patients and clinical users illustrate how helpful they are in practice. The web sites are complemented by the specialist testing that Dr Liehr's own Molecular Cytogenetics section can provide to help characterize ambiguous cytogenetic findings themselves or in collaboration with other centers. As any teacher knows, the best way to learn is to teach others and the Molecular Cytogenetics section runs a series of courses for laboratory and clinical staff so that others can share in their expertise. All this practical experience has been poured into the current volume.

Dr Liehr has always been a collaborative diagnostic and research scientist as can be seen from the wide variety of coauthors on the more than 400 peer-reviewed papers he and his group have published to date. Reflecting his native country's position at the heart of Europe, he has forged strong links with Eastern European and Russian collaborators. He has also collaborated with the Rare Chromosome Support Group (Unique) and contributed to their patient centered publications that are available online (http://www.rarechromo.co.uk/html/home.asp). He has also been involved with publishing as the Editor-in-Chief of the open access journal *Molecular Cytogenetics*, and as an Editorial Board member of several other journals including the *European Journal of Medical Genetics* and the *Journal of Histochemistry & Cytochemistry*.

This book is a comprehensive summary of our state of knowledge at a time of transition when the microscopically visible cytogenetic era is becoming the submicroscopic copy number era. The book will also be useful as the starting point for the many future studies that will become possible through the application of new methods of analysis and imaging. The book also serves as a warning to those who believe that the application of any single technique can provide the answer to any problem and render all past techniques redundant; in contrast, it is often the application of multiple techniques to the same problem that reveals the underlying nature of natural variation that needs to be distinguished from pathological change. This book is not, however, only a source of information; it also packed with practical advice on how best to investigate heterochromatic variation, centromeric variation, and euchromatic variation including unbalanced chromosome abnormalities, euchromatic variants, and supernumerary marker chromosomes. Although the emphasis is on microscopically visible anomalies, submicroscopic copy number variation is also covered where appropriate. The reference section is both extensive and up to date, and much of the information within the book is not easily accessible elsewhere, even with the help of search engines, web browsers, and publication archives.

Thomas Liehr has always been an enthusiastic, responsive, and collaborative colleague. As anyone can see from these words, I would value this book myself, wish the author every success with it, and am glad to recommend it scientists, clinicians, patients, and academics alike.

John C.K. Barber

Honorary Senior Lecture, Department of Human Genetics and
Genomic Medicine, University of Southampton
11th May 2013

ACKNOWLEDGMENTS

This book would not have been possible without all the laboratories sending material to my laboratory. Some of the unpublished data referred to in this book was based on cases sent by Dr. Aktas (Ankara, Turkey), Dr. Bartels (Göttingen, Germany), Dr. Belitz (Berlin, Germany), Dr. Brovko (Kiev, Ukraine), Dr. Cremer (Mannheim, Germany), Dr. Dufke (Tübingen, Germany), Dr. Engels (Bonn, Germany), Dr. Ergul (Kocaeli, Turkey), Dr. Gillessen-Kaesbach (Essen, Germany), Dr. Graf (Hildesheim, Germany), Dr. Hehr (Regensburg, Germany), Dr. Heilbronner (Stuttgart, Germany), Dr. Hexamer-Linder (Hannover, Germany), Dr. Huhle (Leipzig, Germany), Dr. Josic (Vinca, Serbia), Dr. Junge (Dresden/ Erfurth, Germany), Dr. Kistner (Rampe, Germany), Dr. Küchler (Essen, Germany), Dr. Küpferling (Cottbus, Germany), Dr. Lemmens (Aachen, Germany), Dr. Manolakos (Athens, Greece), Dr. Mazauric (Düsseldorf, Germany), Dr. vMehnert (Neu-Ulm, Germany), Dr. Mitter (Leipzig, Germany), Dr. Morlot (Hannover, Germany), Dr. Müsebeck (Bremen, Germany), Dr. Niemann (Overath, Germany), Dr. Pabst (Hannover, Germany), Dr. Petersen (Athens, Greece), Dr. Polityko (Minsk, Belarus), Dr. Sandig (Leipzig, Germany), Dr. Sarri (Athens, Greece), Dr. Schulze (Hannover, Germany), Dr. Seidel (Jena, Germany), Dr. Sheth (Ahmedabad, India), Dr. Simonyan (Yerevan, Armenia), Dr. Steuernagel (Oldenburg, Germany), Dr. Stumm (Berlin, Germany), Dr. Süss (Cottbus, Germany), Dr. Tittelbach (Nürnberg, Germany), Dr. Volleth (Magdeburg, Germany), Dr. Wegner (Berlin, Germany), Dr. Weise (Jena, Germany), Dr. Wieacker (Münster, Germany), and Dr. Yardin (Limoges, France). The multicolor FISH experiments were performed and corresponding figures provided by Monika Ziegler and Katharina Kreskowski (Jena, Germany).

Note: Affiliations are given according to the institutions or cities from which the corresponding persons sent the studied material; they may have since moved.

CHAPTER 1

Introduction

Human chromosomes were first visualized in a microscope in the late 1870s [Arnold, 1879] and were named in 1888 by Heinrich Wilhelm Waldeyer by combining the words *chroma* (Greek χρῶμα, meaning color) and soma (Greek σῶμα, meaning body) [Waldeyer, 1888]. Still, it was another 68 years before the correct modal chromosome number in humans was determined to be 46 [Tijo and Levan, 1956]. Notably, this particular finding was the starting point of the discipline "human cytogenetics," which deals with the numerical and structural analysis of human chromosomes. Since that time in 1956, cytogenetics has played a crucial role in pre- and post-natal, as well as tumor cytogenetic diagnostics and research.

Cytogenetics went through different developmental steps, each providing more and better possibilities for the characterization of structurally abnormal and/or supernumerary chromosomes of unknown origin. The history of human cytogenetics can be divided into three major time periods:

- Prebanding (1879–1970)
- Pure banding (1970–1986)
- Molecular cytogenetic era (1986–today), including the recent invention of array-based comparative genomic hybridization (aCGH)

The identification of the first inborn [Lejeune et al., 1959] and acquired chromosomal abnormalities [Nowell and Hungerford, 1960] occurred in the prebanding era. The banding era started with the development of the Q-banding method by Dr. Lore Zech [Schlegelberger, 2013] in 1968 [Caspersson et al., 1968]. Further techniques like C-banding (CBG) or silver staining of the nucleolus organizing regions (NOR) followed in 1971 and 1976, respectively, and completed the cytogenetic set of standard methods for the next decade [Sumner et al., 1971; Bloom and Goodpasture, 1976]. Currently, the GTG-banding approach (G-bands by Trypsin using Giemsa) [Claussen et al., 2002] is still the gold-standard for all cytogenetic techniques. As a result, translocations, inversions, deletions, and insertions can now be detected and described accurately [Pathak, 1979]. The pure banding era ended in 1986 with the first molecular cytogenetic experiment on human chromosomes [Pinkel et al., 1986]. The preferred technique of

1

molecular cytogenetics is fluorescence in situ hybridization (FISH) [Liehr and Claussen, 2002].

The major proceedings in molecular cytogenetics in the past were the comparative genomic hybridization (CGH) [Kallioniemi et al., 1992] and its array-based variant aCGH [Ren et al., 2005]. Also, multicolor-FISH (mFISH) approaches have been developed since 1989 and continue to this day [Liehr et al., 2013]; for example, a FISH-based detection of copy number variants (CNVs), the so-called parental-origin determination FISH (pod-FISH) approach, was established recently [Weise et al., 2008]. To which extent next-generation sequencing (NGS) approaches can be helpful to study submicroscopic and microscopically visible CNVs remains to be determined. However, aligning or quantification of long repetitive elements present at different loci of the human genome (like satellite III DNA in all acrocentric short arms and homologues regions in 9p12 and 9q13 [Starke et al., 2002]) is no easy task for any kind of sequencing approach [Sipos et al., 2012].

All the aforementioned (molecular) cytogenetic methods (see also Chapter 6) provide information on the human genome at different levels of resolution [Shinawia and Cheung, 2008]. Even so, already an early finding of cytogenetics was that basically no two clinically healthy individuals are alike on a chromosomal level [Ferguson-Smith et al., 1962; Makino et al., 1966]. Thus, cytogenetic visible copy number variations (CG-CNVs) have been detected and characterized since the early days of cytogenetics. Especially prone to formation of CG-CNVs is the constitutive heterochromatin, defined as "regions that are generally late replicating, rich in repetitive DNA sequences, and genetically inert" [Jalal and Ketterling, 2004], and as

that portion of the genome that remains condensed and intensely stained with DNA intercalating dyes throughout the cell cycle. It represents a significant fraction of most eukaryotic genomes and is generally associated with (...) pericentric regions of chromosomes. Contrary to euchromatin, heterochromatic regions consist predominantly of repetitive DNA, including satellite sequences and middle repetitive sequences related to transposable elements and retroviruses. Although not devoid of genes, these regions are typically gene-poor. Establishment of heterochromatin depends on two basic elements: the histone modification code and the interaction of nonhistone chromosomal proteins. [Rizzi et al., 2004]

In addition to heterochromatic CG-CNVs, more recent findings are euchromatic CG-CNVs and submicroscopic ones. Submicroscopic CNVs are detectable only by molecular cytogenetics or aCGH, and thus here abbreviated as MG-CNVs. Also, there is some overlap of MG-CNVs and CG-CNVs [Manvelyan et al., 2011].

1.1. THE PROBLEM

It is well known from banding cytogenetics that some chromosomal regions in the human karyotype are prone to variations more than others; for example, the pericentric regions of chromosomes 1, 9, and 16; the long arm of the Y chromosome; or the short arms of all acrocentrics (Figure 1). More and more regions that might be present in varying copy numbers in the genome without any phenotypic consequences are identified. This accounts

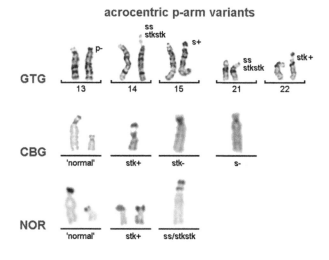

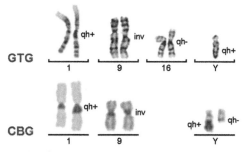

Figure 1 Examples for CG-CNVs as seen in routine cytogenetics; nomenclature is according to ISCN 2013. In the two lines where chromosomes are depicted after GTG-banding the left one of two homologous chromosomes represents the normal variant.

not only for CG-CNVs, but even more for the regions of the genome previously considered as junk DNA (i.e., the MG-CNVs) [Schlattl et al., 2011].

Although CG-CNVs are the focus of this book, MG-CNVs (Figure 2) are also treated here where appropriate and necessary. In banding-cytogenetics the main clinical problem is to differentiate similar-looking benign from pathological CG-CNVs, but for MG-CNVs it is often unclear which are benign and which are not [Girirajan et al., 2011]. To obtain up-to-date information for MG-CNV databases see Chapter 7, section 7.1.

1.1.1. Definition of CG-CNVs versus MG-CNVs

CG-CNVs are defined as gain or loss of genetic material that leads to a different appearance of a chromosome visualized in banding cytogenetics. MG-CNVs are detectable in aCGH together with a normal karyotype.

For a more comprehensive characterization of a CG-CNV or an MG-CNV, molecular cytogenetics is applied.

Note: Except for size, MG-CNVs and CG-CNVs do not have a natural/biological difference.

CG-CNVs comprise enlargement or reduction in the size of chromosomal subbands and/or appearance of additional band(s). An MG-CNV can mean an amplification or loss of repetitive DNA stretches. One of the first identified MG-CNVs is a 28 kb deletion in 15q11.2 [Buiting et al., 1999] (Figure 2). There are inversions on the microscopic level also, which are considered as polymorphisms in this book. Gardner et al. (2011) also treat fragile sites [Mrasek et al., 2010] as chromosomal variants, which are not incorporated here. Finally, there are many other polymorphic variants in the human genome, which concern a few bases only, like single nucleotide

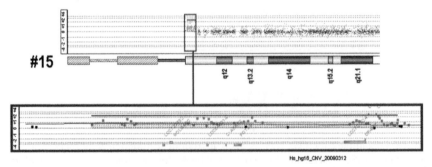

Figure 2 A common MG-CNV is located in 15q11.2, as detected here in a normal person by aCGH as a gain of copy numbers.

polymorphisms (SNPs) [Weiss, 1998] or so-called insertion/deletion (indel) variants (http://www.hgvs.org/mutnomen/recs-DNA.html#DNA; not topics in this book). It goes without saying that any kind of CNV may be found in otherwise normal as well as abnormal karyotypes (see also sections 1.1.2 and 1.3.1)

1.1.2. CG-CNVs without Clinical Consequences

There are numerous regions within the human genome that are considered CG-CNVs without any phenotypic consequences (see Chapters 2 and 5). However, CG-CNVs can be mixed up with chromosomal imbalances, which may be adverse and lead to clinical drawbacks for its carriers. Still, known benign and pathological chromosomal imbalances in many cases can be easily distinguished by the corresponding FISH probes. As already mentioned, there are also euchromatic CG-CNVs in addition to heterochromatic ones (see Chapter 5).

1.1.2.1. Heterochromatic CG-CNVs

Heterochromatic CG-CNVs without clinical consequences can be found most often as size variations of the centromeric satellite DNA of any of the 24 human chromosomes; the short arms of the acrocentric chromosomes 13–15 and 21–22; the pericentric heterochromatin of chromosomes 1, 9, and 16; and the Y-chromosomal band q11.2 (Figure 3). In the 1970s, fluorescence approaches like DAPI/DA (4,6-diamino-2phenyl-indole/distamycin A) staining were widely used to subdifferentiate these heteromorphic patterns [Jalal et al., 1974; Schwanitz, 1976]. Another impressive idea was to combine restriction endonucleases with GTG-banding [Babu et al., 1988]. However, all these studies went completely out of style after FISH was introduced. Still, DAPI-banding pattern might be considered in FISH in some special cases as well (see the Color Plates).

Cytogenetically visible heterochromatic blocks may be inserted at any place in the genome [Bucksch et al., 2012]. Also, exclusively heterochromatic extra-chromosomes, so-called small supernumerary marker chromosomes (sSMC), can be considered as heterochromatic CG-CNVs without clinical consequences [Liehr, 2012].

1.1.2.2. Euchromatic CG-CNVs

In the majority of cases, euchromatic imbalances of cytogenetically visible size (i.e., several megabasepairs (Mb)) lead to severe clinical consequences [Schinzel, 2001]. Nonetheless, in recent years more and more so-called

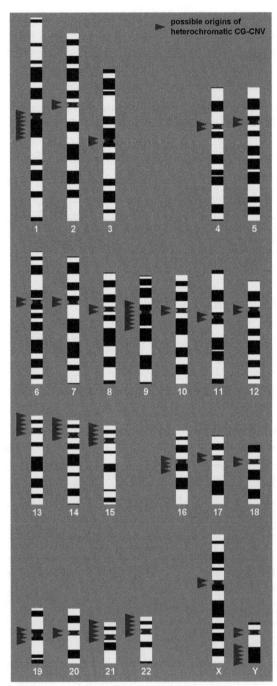

Figure 3 Heterochromatic CG-CNVs may derive from the regions marked by arrowheads.

euchromatic variants (EV; i.e., large-scale cytogenetically visible copy number variants) and unbalanced chromosome abnormalities without phenotypic consequences (UBCA) have been reported [Barber, 2005 and http://www.ngrl.org.uk/Wessex/collection/index.htm]. These CG-CNVs can be due to gain or loss of copy numbers at the original location; moreover such regions can be translocated to another chromosomal region due to rearrangements. It is suggested that any region not containing dose-dependant genes may be amplified or lost in the genome without clinical consequences, as also seen in euchromatic sSMC [Liehr, 2012].

In 1993 it was nicely summarized [Berg et al., 1993] that structural autosomal imbalances in typical cases may lead to syndromes with a complex of minor anomalies and congenital malformations. The latter "suggests the importance of gene interaction in determining the phenotypic picture of autosomal imbalance syndromes" [Berg et al., 1993]. Duplication-related syndromes are much more frequent than deletion-related ones, and thus it is common sense that in general, duplications of several Mb in size are better tolerated by the human genome than deletions of the same sizes. This has also recently been confirmed on the level of microduplications and micro-deletions [Roa and Lupski, 1994; Weise et al., 2012]. Overall, chromosomal imbalances can point toward dosage-sensitive genes, being responsible for specific syndromes or clinical features. However, even though less frequent, there are also genetically relevant regions that can be tolerated even with three or more copies in the human genome. In summary, studying carriers of specific chromosomal imbalances including euchromatic CG-CNVs can contribute to genotype–phenotype correlations, and give hints where copy-number-(in)sensitive genes are located in our genome.

1.2. FREQUENCY AND CHROMOSOMAL ORIGIN OF CYTOGENETICALLY VISIBLE COPY NUMBER VARIANTS (CG-CNVs) WITHOUT CLINICAL CONSEQUENCES

Basically no two clinically healthy individuals are really the same on a chromosomal level [Ferguson-Smith et al., 1962; Makino et al., 1966] as they aren't on a molecular genetic level [1000 Genomes Project Consortium et al., 2012]. Thus CG-CNVs can be expected to be present in every individual. The majority of size and structural differences are found in 1q12, 9q12, 13pter-q11, 14pter-q11.1, 15pter-q11.1, 16q11.2, 19p12-q12, 21pter-q11.1, 22pter-q11.1 and in male in Yq12 (Figure 3); further chromosomal origins of CG-CNV are specified in Chapters 2 and 5.

Applying the cytogenetic approaches from the mid 1970s, an average of four to five chromosomal heteromorphisms per person was found [Geraedts and Pearson, 1974; McKenzie and Lubs, 1975]. Now this is accentuated by the findings of aCGH, showing an average of 0.5 Mb MG-CNVs difference in each individual [Girirajan et al., 2011]. Thus, a unique combination of CG-CNVs and MG-CNVs is present in every human being, except possibly for monozygote twins, even though distinct MG-CNVs were already found there as well [Bruder et al., 2008]. Still, it is possible to provide approximate incidences of subgroups of CG-CNVs, as summarized in Table 1, even though the observed frequencies varied according to the applied cytogenetic methods, criteria of observation, and ethnic background [Gardner et al., 2011].

1.3. PRACTICAL MEANING OF CG-CNVs IN DIAGNOSTICS AND RESEARCH

Currently, CG-CNVs mainly are considered a diagnostic problem to be resolved; benign CG-CNVs shall be distinguished from pathological chromosomal imbalances. Nonetheless, in exceptional cases a CG-CNV can be useful in the following instances [Schinzel 1982]:
- To determine paternity
- To differentiate between mono- and dizygote twins
- To determine parental origin of additional and/or derivative chromosomes or of haploid sets in triploidy and chimeras
- To detect maternal contamination in amniotic fluid cell cultures
- To follow up bone marrow transplantation
- To find genetic linkage [Ferguson-Smith et al., 1975]

Until now none of the many suggested different correlations of CG-CNVs with any clinical phenotype has been persistent to another check-up. This is especially true for tumors (see Chapter 3, section 3.2.1), infertility [Kosyakova et al., 2013], and many other clinical features [Starke et al., 2002].

1.3.1. Multiple CG-CNVs

CG-CNVs are always present together with MG-CNVs [Girirajan et al., 2011]. Also, two or more CG-CNVs can regularly be found in chromosomal analysis (section 1.2). Interestingly, "it was recently observed that *more than one (submicroscopic) CNV* (larger than 500 kb) can contribute to severe developmental delay and often is responsible for phenotypic variability

Table 1 Frequencies of CG-CNVs and Inversion-Variants of Chromosomes 2 and 9

Chromosomal origin	Number of cases included and corresponding frequency	References
1qh+	98/39, 612 0.25%	Walzer et al., 1969 Bochkov et al., 1974 Nielsen, Friedrich, Areidarsson, 1974 Tüür et al., 1974 Mikelsaar et al., 1975 Hamerton et al., 1975 Nielsen and Sillesen, 1975 Schwanitz, 1976
9qh+	87/26, 143 0.33%	Walzer et al., 1969 Bochkov et al., 1974 Nielsen, Friedrich, Areidarsson, Zeuthen, 1974 Mikelsaar et al., 1975 Nielsen and Sillesen, 1975 Schwanitz, 1976 Metaxotou et al., 1978
p+ (enlargement of acrocentric short arms 13, 14, 15, 21 or 22)	962/40, 397 2.38%	Walzer et al., 1969 Bochkov et al., 1974 Lubs and Ruddle, 1970 Zankl and Zang, 1971 Hamerton et al., 1972 Mikelsaar et al., 1973 Hamerton et al., 1975 Nielsen and Sillesen, 1975 Schwanitz, 1976
p− (shortage of acrocentric short arms 13, 14, 15, 21 or 22)	34/30, 117 0.11%	Walzer et al., 1969 Bochkov et al., 1974 Hamerton et al., 1975 Nielsen and Sillesen, 1975
16qh+	149/39, 773 0.37%	Walzer et al., 1969 Bochkov et al., 1974 Lubs and Ruddle, 1970 Zankl and Zang, 1971 Hamerton et al., 1972 Hamerton et al., 1975 Nielsen and Sillesen, 1975 Mikelsaar et al., 1973

(Continued)

Table 1 Frequencies of CG-CNVs and Inversion-Variants of Chromosomes 2 and 9—cont'd

Chromosomal origin	Number of cases included and corresponding frequency	References
Yqh+	293/37, 373 0.78%	Bochkov et al., 1974 Lubs and Ruddle, 1970 Zankl and Zang, 1971 Hamerton et al., 1972 Hamerton et al., 1975 Mikelsaar et al., 1973 Nielsen and Sillesen, 1975
Yqh−	33/37, 373 0.09%	Bochkov et al., 1974 Hamerton et al., 1975 Nielsen and Sillesen, 1975
47,XN,+mar	$2\times10^6/7\times10^9$ 0.03%	Liehr, 2012
inv(2)(p11.2q13)	19/16, 533 0.11%	MacDonald and Cox, 1985 Djalali et al., 1986 Vejerslev and Friedrich, 1984
inv(9)(p11q13)	173/6, 050 2.86%	Craig-Holmes et al., 1973 Mutton and Daker, 1973 de la Chapelle et al., 1974 Hansmann, 1976 Metaxotou et al., 1978 McKenzie and Lubs, 1975 Verma et al., 1981 Yamada, 1992 Uehara et al., 1992 Ait-Allah et al., 1997

associated with genomic disorders" [Girirajan et al., 2011]. This phenomenon is called a "two-hit" model [Girirajan et al., 2010] and has not yet been tested for patients with multiple CG-CNVs or with CG-CNVs and MG-CNVs and syndrome or disease.

1.4. SUBMICROSCOPIC CNVs (MG-CNVs)

MG-CNVs range from 1 kilobasepair (kb) to several Mb in size, are distributed along the entire human genome, and with increasing resolution of aCGH platforms, more and more such variants are detected. These structural variants show variable copy numbers when compared to a reference genome and may include deletions and duplications of genomic loci [Feuk et al., 2006]. They may encompass as much as 12% of the human genome [Redon et al., 2006]. Some of these aberrations are apparently benign MG-CNVs and are usually inherited from a parent [Lee et al., 2007]. When determined as de novo, genomic imbalances are considered more likely pathological [Tyson et al., 2005]. It is also known that any individual carries about a thousand MG-CNVs ranging from only a few hundred basepairs (bp) to over 1 Mb [Conrad et al., 2010]. This can lead to either too many or too few copies of dosage-sensitive genes, which might result in phenotypic variability, complex behavioral traits, disease susceptibility, and predispositions such as HIV, adipositas, and psoriasis [Canales and Walz, 2011].

Identical MG-CNVs may be found in a healthy person and his or her diseased offspring. Here either the two-hit model (section 1.3.1) may be helpful for explanation, or other molecular mechanisms by which rearrangements of the genome may convey or alter a disease phenotype could be involved; for example, the unmasking of either a recessive mutation or a functional polymorphism of the remaining allele could be disease-causing when a deletion occurs [Lupski and Stankiewicz, 2005; Kurotaki et al., 2005; Albers et al., 2012]. Finally, some biological phenomenon and even diseases were recently correlated with higher order repeat sizes, like D4Z4 in 4q35 with facioscapulohumeral muscular dystrophy [Lemmers et al., 2012] or DXZ4 with a function in X chromosome inactivation [Warburton et al., 2008].

CG-CNVs: What Is the Norm?

Because this book deals with CG-CNVs (i.e., deviations from the norm), we first need to define what the norm is—this is quite tricky, as outlined in this chapter.

2.1. ACROCENTRIC CHROMOSOMES' SHORT ARM VARIANTS

The five human acrocentric chromosomes are numbered 13, 14, 15, 21, and 22. They all have a cytogenetically similar short arm that is extremely gene-poor. Their main contribution for the cell is that the acrocentric short arms are carriers of the nucleolus organizing regions (NOR) in subbands p12.

Apart from one study performed by fluorescence staining [Verma et al., 1977], it has never been defined what the typical size or length of an acrocentric short arm is. Hints may come from idiograms presented in the different versions of the international system for human cytogenetic nomenclature (ISCN) and the genome browsers (e.g., see http://genome.ucsc.edu/index.html).

A closer look at the acrocentric short arms in the standard idiograms of ISCN 1978, ISCN 1985, ISCN 2005, and ISCN 2009 reveals quite big differences between these four versions; in ISCN 2013 the idiograms were not changed. For unknown reasons there is a tendency to reduce the normal size of an acrocentric short arm from ISCN version to version (Table 2).

Also, molecular sizes given in the genome browsers alter between NCBI34/hg16 toward the version GRCh37/hg19 (Table 3). Size differences of up to 3.7 Mb are present here. The estimated size of the heterochromatic and acrocentric portions of the human genome that still are not sequenced is estimated to be approximately 200 Mb [She et al., 2004].

A suggestion for normal/average length of acrocentric short arms (p-arms) was given by Verma et al. (1977), who recommended using the length of chromosome 18p of the same metaphase spread as orientation (Table 2, last column; Figure 4A) and defining significant enlargement or loss as CG-CNVs (see Chapter 5, section 5.1.1).

Table 2 Length of Short Arms with Respect to Each Individual Acrocentric Chromosome Long Arm Size (length of p-arm: length of q-arm) According to ISCN 1978, ISCN 1985, ISCN 2005, and ISCN 2009/2013

Chromosome position (cytoband)	ISCN 1978	ISCN 1985p63-64	ISCN 2005	ISCN 2009 ISCN 2013	Average length (i.e., chromosome 18p)
13pter–p11.1	0.32	0.22	0.21	0.17	0.16
14pter–p11.1	0.30	0.24	0.24	0.18	0.17
15pter–p11.1	0.30	0.24	0.25	0.19	0.18
21pter–p11.1	0.71	0.56	0.57	0.48	0.41
22pter–p11.1	0.63	0.44	0.46	0.40	0.39

2.2. VARIANTS OF THE CENTROMERIC REGIONS

"The centromere, recognized cytologically as the primary constriction, is essential for chromosomal attachment to the spindle and for proper segregation of mitotic and meiotic chromosomes" [Sullivan et al., 1996]. It is common sense that a normal centromere, necessary for the formation of the kinetochor, contains α-satellite DNA stretches of several kb to Mb in size. However, defining a norm for centromeric size is hindered by at least two nomenclatorial problems. The first one is present on cytoband level; that is, the centromeric regions are described in a nonuniform way: they are reported to span subbands p11-q11 (chromosomes 4, 19), p11-q11.1 (chromosomes 3, 5), p11.1-q11 (chromosomes 1, 9, 11-13), or p11.1-q11.1 (chromosomes 2, 6-8, 10, 14-18, 20-22, X, Y). Centromeric bands cover in the idiograms 2.5% (chromosome 1) up to 9% (Y chromosome) of the corresponding whole chromosome length in ISCN 2013.

The second problem in finding a standard for centromeric size concerns the molecular level. Similar to the discussion in section 2.1 regarding acrocentric short arms, there is no common definition of what the "normal size" of a human centromeric/(α-)satellite region is. The confusion about this point is reflected best in the human genome browsers (e.g., http://genome.ucsc.edu/index.html) and their different versions (Table 4). As an example, the size for centromeric region of chromosome 17 varies between hg16 and hg19 from 3.5 Mb to 0.7 Mb and 1.1 Mb to 3.6 Mb.

Note: All these version-specific changes of centromere positioning may also lead to huge mapping differences of genes, especially in the long arm (q-arm) of the chromosomes.

Table 3 The Acrocentric Short Arm Positions and Sizes in Different Versions of NCBI Versions 34 to 36 and GRCh37 (NCBI34/hg16 to GRCh37/hg19)

Chromosome position (cytoband)	NCBI34/hg16	NCBI35/hg17	NCBI36/hg18	GRCh37/hg19	Average length (i.e., chromosome 18p)
13pter-p11.1	1–12,600,001	1–13,500,001	like previous version	1–16,300,001	1–15,400,000
14pter-p11.1	1–13,600,001	like previous version	like previous version	1–16,100,001	1–15,400,000
15pter-p11.1	1–14,100,001	like previous version	like previous version	1–15,800,001	1–15,400,000
21pter-p11.1	1–10,100,001	like previous version	like previous version	1–10,900,001	1–15,400,000
22pter-p11.1	1–9,600,001	like previous version	like previous version	1–12,200,001	1–15,400,000

Table 4 The Centromeric Positions and Sizes in Different Versions of NCBI Versions 34 to 36 and GRCh37 (NCBI34/hg16 to GRCh37/hg19)

Chromosome position (cytoband)	NCBI34/hg16	NCBI35/hg17	NCBI36/hg18	GRCh37/hg19
1p11.1-q11	120,100,000–126,100,000	121,100,001–127,900,000	121,100,001–128,000,000	121,500,001–128,900,000
2p11.1-q11.1	91,400,001–96,300,000	91,000,001–95,800,000	91,000,001–95,700,000	90,500,001–96,800,000
3p11.1-q11.1	87,900,001–93,100,000	87,900,001–93,200,000	89,400,001–93,200,000	87,900,001–93,900,000
4p11-q11	48,900,001–53,200,000	like previous version	48,700,001–52,400,000	48,200,001–52,700,000
5p11-q11.1	45,800,001–50,500,000	like previous version	like previous version	46,100,001–50,700,000
6p11.1-q11.1	58,700,001–63,300,000	58,800,001–63,400,000	58,400,001–63,400,000	58,700,001–63,300,000
7p11.1-q11.1	56,600,001–61,200,000	56,900,001–61,200,000	57,400,001–61,100,000	58,000,001–61,700,000
8p11.1-q11.1	43,100,001–48,000,000	42,100,001–48,100,000	43,200,001–48,100,000	43,100,001–48,100,000
9p11.1-q11	44,300,001–57,600,000	45,900,001–59,200,000	46,700,001–60,300,000	47,300,001–50,700,000
10p11.1-q11.1	38,300,001–41,800,000	38,800,001–41,900,000	38,800,001–42,100,000	38,000,001–42,300,000
11p11.11-q11	51,600,001–56,700,000	51,400,001–56,700,000	51,400,001–56,400,000	51,600,001–55,700,000
12p11.1-q11	33,200,001–36,500,000	like previous version	like previous version	33,300,001–38,200,000
13p11.1-q11	12,600,001–17,300,000	13,500,001–18,400,000	like previous version	16,300,001–19,500,000
14p11.1-q11.1	13,600,001–18,000,000	13,600,001–19,000,000	13,600,001–19,100,000	16,100,001–19,100,000
15p11.1-q11.1	14,100,001–18,200,000	14,100,001–18,300,000	14,100,001–18,400,000	15,800,001–20,700,000
16p11.1-q11.1	35,700,001–41,900,000	34,400,001–40,700,000	like previous version	34,600,001–38,600,000
17p11.1-q11.1	22,400,001–25,900,000	22,100,001–22,800,000	22,100,001–23,200,000	22,200,001–25,800,000
18p11.1-q11.1	15,300,001–17,300,000	15,400,001–17,300,000	like previous version	15,400,001–19,000,000
19p11-q11	26,700,001–30,200,000	like previous version	like previous version	24,400,001–28,600,000
20p11.1-q11.1	25,800,001–29,700,000	25,700,001–28,400,000	like previous version	25,600,001–29,400,000
21p11.1-q11.1	10,100,001–13,200,000	10,000,001–13,200,000	like previous version	10,900,001–14,300,000
22p11.1-q11.1	9,600,001–16,300,000	like previous version	like previous version	12,200,001–17,900,000
Xp11.1-q11.1	55,600,001–60,600,000	56,500,001–61,500,000	56,600,001–65,000,000	58,100,001–60,600,000
Yp11.1-q11.1	9,700,001–12,800,000	11,200,001–12,400,000	11,200,001–12,500,000	11,600,001–13,400,000

On the sequence level there are centromeric gaps at almost every proximal p-arm and q-arm, as the centromeric regions "have been largely excluded from (...) sequencing efforts due in part to the highly repetitive DNA content" [Schueler et al., 2001]. Thus, the centromeric size in the genome browsers is more or less arbitrarily set, as the relationship between physical and genetic distances is unknown [Rudd et al., 2006]. In 2004, an estimated 10.5 Mb remained to be sequenced within the pericentric regions of the human genome [She et al., 2004].

A direct estimation of genetic distances with respect to the centromere was done for only a few chromosomes [She et al., 2004]. Still valid is the following statement from 1999:

> The absence of knowledge of the genetic and physical structures of the centromeric regions is (...) due to particular traits of the centromeric regions, their poor, if any, representation in the genetic maps produced so far and the total lack of sequence-tagged sites within them. [Puechberty et al., 1999]

Human alphoid DNA detected in molecular cytogenetics consists of 171 bp repeat units, which are basically homologous to each other. Nonetheless, almost every human chromosome pair has an individual repeat differing from the others by 20 to 40% [Paar et al., 2007]. Exceptions are the following groups of chromosomes, sharing identical α-satellite-sequences: chromosomes 1, 5, and 19 (D1Z1/ D5Z2/ D19Z3); chromosomes 13 and 21 (D13Z1/ D21Z1); chromosomes 14 and 22 (D14Z1/ D22Z1) [Jørgensen et al., 1997]; chromosomes 2 and 20 (probe pBS4D) [Rocchi et al., 1990]; chromosomes 5 and 19 (D5Z1/ D19Z2) [Hulsebos et al, 1988]; and chromosomes 4 and 9 (probe p4n1/4 [D'Aiuto et al. 1997] and probe pG-Xba11/340 [Hulsebos et al, 1988]). The 171 bp repeats are present as hundreds to thousands of head-to-tail tandem copies at the centromere, in most cases are arranged as so-called higher-order repeat units of several kb in size [Lo et al., 1999]. The centromeric regions are "highly polymorphic and constitute rich (...) source of DNA-based variation in the human genome" [Willard et al., 1987]. α-satellite DNA represents 3 to 4% of the chromosomal DNA [Lo et al., 1999]; however, parts of these stretches are located outside of the centromeric regions and normally are not arranged in higher order repeats (HORs) [Warburton et al., 2008]. According to Marzais et al. (1999) alphoid DNA on one chromosome may span from about 250 kb to over 4,000 kb (see Chapter 5, section 5.1.2). It was also suggested in a single study that the size of the alphoid DNA might be proportional to the size of the chromosome [Sánchez et al., 1991].

A good summary of this section may be the following:

Although centromeres are unambiguously defined at the cytogenetic level as primary constrictions of chromosomes, their definition is not as clear at the molecular DNA level, as the DNA sequences necessary and sufficient to ensure proper centromere function in both mitosis and meiosis in higher eukaryotes remain unknown. A certain amount of α-satellite DNA is, however, always detected at primary constrictions and can, therefore, be considered as spanning centromeres in all human chromosomes (excluding neo-centromeres). Other satellite DNAs and sometimes other alphoid sequence subsets map close to the primary constrictions on the proximal p- or q-arms, a situation that is highly variable among the chromosomes and the satellite DNAs considered. [Puechberty et al., 1999]

CG-CNVs of the centromeric regions were described for all human chromosomes applying banding, as for molecular cytogenetics. For modes of their formation see Chapter 4, section 4.2; chromosome specific details may be found in Chapter 5, section 5.1.2.

2.3. VARIANTS OF NONCENTROMERIC HETEROCHROMATIN

Human noncentromeric heterochromatin can be found specifically in regions 1q12, 9q12, 16q11.2, and Yq12. Although in the different ISCN versions [ISCN 1978, 1985, 2005, 2009/2013] the size of these regions remains relatively stable, their sizes and positions in the genome browsers change in the same interesting way as those of the short acrocentric chromosome arms (Table 3) and those of the centromeres (Table 4), as highlighted in Table 5.

It has previously been suggested to use the short arms of chromosome 16 of the same metaphase as a reference size for noncentromeric heterochromatin [McKenzie and Lubs, 1974]. However, as Figure 4B shows, this is only helpful to define a maximum size of these heterochromatic blocks; that is, all corresponding bands larger than 16p would be a qh+ variant. In Chapter 5, section 5.1.2.9, a FISH-based definition of 9qh+ (Figure 16) and the definition of the other three variants (Chapter 5, sections 5.1.2.1, 5.1.2.16, and 5.1.2.24) is suggested and refined.

2.4. UNBALANCED CHROMOSOME ABNORMALITIES (UBCAs) WITHOUT CLINICAL CONSEQUENCES

A UBCA without clinical consequences is considered here as a euchromatic form of a CG-CNV. The norm in connection with a UBCA without clinical consequences is easy to settle: it is a numerically and structurally

Table 5 The Positions of Heterochromatic Blocks in Different Versions of NCBI Versions 34 to 36 and GRCh37 (NCBI34/hg16 to GRCh37/hg19)

Chromosome position (cytoband)	NCBI34/hg16	NCBI35/hg17	NCBI36/hg18	GRCh37/hg19
1q12	126,100,001–140,800,00	127,900,001–141,600,00	128,000,001–142,400,00	128,900,001–142,600,00
9q12	57,600,001–62,900,000	59,200,001–67,500,000	60,300,001–70,000,000	50,700,001–65,900,000
16q11.2	41,900,001–46,800,000	40,700,001–45,500,000	like previous version	38,600,001–47,000,000
Yq12	27,700,001–50,286,555	27,100,001–57,701,691	27,200,001–57,772,954	28,800,001–59,373,566

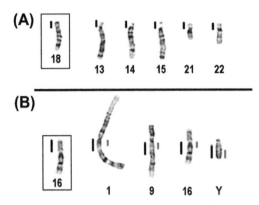

Figure 4 In literature suggestions were made for an average length of acrocentric short arms and also what should be considered an abnormal length of 1q12, 9q12, 16q11.2, and Yq12. **A.** Average length of all acrocentric short arms should be about the size of the short arm of a chromosome 18 of the same metaphase spread; here an example of one chromosome 18 used as standard (black bar on left side in the black square) and one chromosome 13, 14, 15, 21, and 22, each with short arms of a normal length. **B.** Average length of heterochromatic blocks in 1q12, 9q12, 16q11.2, and Yq12 should not exceed the length of the short arm of chromosome 16 of the same metaphase spread (black bar on left side in the black square). The length of the four corresponding regions on chromosomes 1, 9, 16, and Y were considerably smaller than this size (highlighted by gray bars on the right side of each chromosome); however, they would not be designated as smaller than average.

normal human karyotype, especially in terms of the euchromatic chromosomal regions. UBCAs would be assumed if a deviation from 46 chromosomes or normal banding patterns is observed. UBCAs without clinical consequences can be best characterized as such if they are present in more than one individual within one family over more than one generation. The first UBCA was observed in banding-cytogenetics in 1973 as a der(6) t(6;21)(p2?5;q11) [Borgaonkar et al., 1973], since surprisingly, this chromosomal imbalance did not lead to any clinical consequences for its carrier.

A UBCA presents like an adverse chromosomal aberration and involves euchromatic chromosomal bands. It can be present as a derivative chromosome, an insertion, a duplication, or a deletion, leading to a cytogenetically visible gain or loss of genetic material. In general, a single UBCA without clinical consequences cannot be determined on the pure cytogenetic level to be deleterious for its carrier. Just a comparison with literature and family studies can be indicative here, especially for prenatal cases (see Chapter 5, section 5.2).

2.5. SMALL SUPERNUMERARY MARKER CHROMOSOMES (sSMCs)

An sSMC is cytogenetically definable as a structurally abnormal chromosome that cannot be identified or characterized unambiguously by banding cytogenetics alone, and is (in general) equal in size or smaller than a chromosome 20 of the same metaphase spread. sSMCs can also be present (1) in a karyotype of 46 normal chromosomes, (2) in a numerically abnormal karyotype (e.g., Turner or Down syndrome), or (3) in a structurally abnormal but balanced karyotype (e.g., Robertsonian translocation) or ring chromosome formation. An sSMC, as a UBCA, should not be observed in a normal human karyotype. If present, sSMCs may be pure heterochromatic but also consist of euchromatic parts. Interestingly, both conditions may be non-deleterious for the individual but also associated with a clinical phenotype. Indeed, sSMCs are without clinical consequences for 70% of their carriers [Liehr, 2012] and thus can be considered as CG-CNVs as well. Also, they can be regarded as special kinds of UBCA without clinical consequences.

2.6. EUCHROMATIC VARIANTS (EVs)

EV-regions are those that may contain stretches with (hemi)heterochromatic and euchromatic features. In other words, this kind of CG-CNV resembles duplications and contains "genes and pseudogenes which are polymorphic in the normal population and only reach the cytogenetically detectable level, when multiple copies are present. These EVs segregate in most families without apparent phenotypic consequences" [Barber, 2005]. The definition of CG-CNV and MG-CNV is especially valid here—besides the size, which may hamper or enable cytogenetic based detection, there is often no real difference between these two groups of CNVs (see Chapter 1, section 1.1.1).

The first EV was reported in 1980 as a 9ph+ variant [Buckton et al., 1980] (see Chapter 5, section 5.2.1.1). Currently the following regions are considered to be able to form an EV: 4p16.1, 8p23.1, 9p12/9p11.2-p13, 9q12-q21.12/9q13-q21.12, 15q11.2, and 16p11.2 (Figure 5) [Barber, 2005], even though others may be considered as such as well (Chapter 5). Size variations of those bands are indicative for the presence of an EV. The euchromatic bands are defined in all ISCN versions uniformly [ISCN 1978, 1985, 2005, 2009/2013]. Overall, an EV is basically nothing other than a MG-CNV, which in some instances is so large that it becomes a CG-CNV.

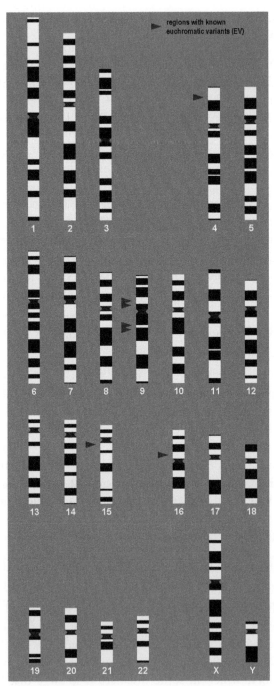

Figure 5 Euchromatic variants (EV) may derive from the regions marked by arrowheads.

2.7. GONOSOMAL DERIVED CHROMATIN

2.7.1. Gonosomal Derived Heterochromatin

Apart from the centromeric region of the X- and Y-chromosomes (see section 2.2) there is a large heterochromatic block only in Yq12. As highlighted in section 2.3 and Chapter 5, sections 5.1.1.1.2.2 through 5.1.1.1.2.4, and 5.1.3.1, this part can be enlarged, diminished, and also translocated to other chromosomes. For possible modes of formation see Chapter 4, sections 4.1 and 4.2.

2.7.2. Gonosomal Derived Euchromatin

2.7.2.1. Y-Chromosome

It is well known that additional copies of an entire Y-chromosome (karyotype 47,XYY) have no, or only minor, effects on the male phenotype [Stochholm et al., 2012]. While unbalanced translocations of Y-chromosomal euchromatin to another human chromosome are rare [Manvelyan et al., 2007, case 30] isochromosome formation of (almost) the entire Y-chromosome are reported relatively often. Since in those instances the dicentric Y-chromosome tends to be lost during cell division, the majority of such cases is detected clinically as a Turner syndrome mosaic, with karyotypes, for example, like mos 45,X/46,X,der(Y)(qter->p11.32::p11.32->qter) [Liehr et al., 2007]. Overall, euchromatic Y-chromosome duplications have by far no more severe an impact than gain of copy numbers of autosomal euchromatin. Also for the individual, loss of euchromatic Y-chromosome material is less problematic; indeed, a monosomy X due to loss of a Y- or an X-chromosome is the only viable monosomic condition in humans. It may lead to the variably expressed Turner syndrome, most often going along with infertility.

Overall, gain or loss of euchromatic Y-chromosome material might carefully be considered as a CG-CNV as well.

2.7.2.2. X-Chromosome

For the X-chromosome, the Lyon-hypothesis has to be observed. Every human, even every mammalian female cell, inactivates transcription in one of both X-chromosomes to balance X linked gene dosage between male and female. In male and female all (additional) X-chromosomes except one undergo X-chromosome inactivation. This epigenetic event leads to gene silencing along almost the entire X-chromosome. But not all genes on the inactive X-chromosome are inactivated: genes in the pseudo-autosomal regions, the regions of the X-chromosome homologous to the Y-chromosome and responsible for XY-pairing during meiosis, as well as 15 to 20% of individual genes on the X-chromosome are not inactivated; another 10% of the genes escape inactivation partially. X-chromosome inactivation needs for its initiation the so-called X-inactivation-specific transcript (XIST) gene in Xq13.2 [Pontier and Gribnau, 2011].

Balanced and unbalanced translocations involving the X-chromosome including the XIST-gene region need to be checked carefully for their clinical relevance, since X-inactivation may also spread to genetic regions not derived originally from the X-chromosome itself. Thus, in large part and otherwise not viable, UBCA without severe clinical consequences may be observed [Stankiewicz et al., 2006]. Accordingly, even though additional copies of an X-chromosome may not be considered as CG-CNV, even large unbalanced translocations involving the XIST-gene may behave exactly like this.

For loss of parts or of an entire X-chromosome, what was stated in section 2.7.2.1 for loss of euchromatic Y-chromosome material is valid. (Partial) monosomy of the X-chromosome is normally associated with features of Turner syndrome.

2.8. MG-CNVs

The DNA sequences obtained in the Human Genome [Lander et al., 2001] and Celera Genomics [Venter et al., 2001] projects serve as reference for MG-CNVs. In other words, the DNA of about 10 different healthy, randomly selected persons are used as reference for gain or loss of MG-CNV [http://en.wikipedia.org/wiki/Human_Genome_Project]. Thus, presently it is not known what a normal or average variant is, and what real gain or loss compared to the majority of the population is. To solve this problem, the 1000 Genomes Project is a first step [1000 Genomes Project Consortium, 2012].

CHAPTER 3

Inheritance of CG-CNVs

3.1. FAMILIAL CG-CNVs

The great majority of the CG-CNVs detected in routine chromosome analysis can be found in one of the parents of the studied individual. This holds especially true for the heterochromatic CG-CNVs. In exceptional cases it is even possible to track a CG-CNV for more than 10 generations [Genest, 1972; Soudek, 1973], and a Mendelian inheritance pattern was shown for them [Carnevale et al., 1976; Verma et al., 1977] (Figure 6). Since case reports on de novo CG-CNVs are very limited (see section 3.2), it may be speculated that most CG-CNVs seen in diagnostics have been present in the human population for many generations.

Interestingly, CG-CNVs are much more frequently observed in individuals of African origin than in other human populations [Lubs and Ruddle, 1971, Hsu et al., 1987]. This finding pairs well with the recent discovery that "substructured populations of Africa retain an exceptional number of unique variants, and there is a dramatic reduction in genetic diversity within populations living outside of Africa" [Henn et al., 2012].

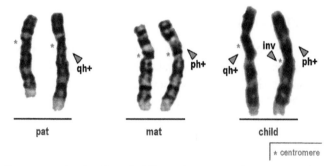

Figure 6 Two-generation family with different heteromorphic patterns of chromosome 9: the father has a normal chromosome 9 and one with a 9qh+, and the mother has one with a 9ph+ variant. The unborn child inherited both variant chromosomes 9, and additionally acquired a pericentric inversion in the maternal-derived chromosome 9. Banding-cytogenetic results (GTG) were substantiated by C-banding (CBG) and FISH as well (results not shown).

Euchromatic CG-CNVs were discovered later than heterochromatic ones; unfortunately no population studies are available for them. However, it seems that most of them behave like heterochromatic ones (see Chapter 5).

3.1.1. Familial CG-CNVs and Infertility

In the early days of banding cytogenetics (from 1960 to the 1990s) there were numerous studies concerning CG-CNVs [Schwanitz, 1976]. Even though some of them seemed initially to be indicative for a clinical meaning of CG-CNVs [Adhvaryu and Rawal, 1991; Starke et al., 2002] none of those suggestions could be substantiated later. Nowadays it is common sense that heterochromatic CG-CNVs are just polymorphisms without any clinical meaning and are inherited even in large pedigrees in a Mendelian fashion (see section 3.1).

Nonetheless, it is thought that heteromorphic chromosomes are more likely to be detected in patients studied for infertility than in the general population [Schwanitz, 1976; Codina-Pascual et al., 1976; Chatzimeletiou et al., 2006], even though this was already disproven convincingly in a study in 1982 [Blumberg et al., 1982]. Even more confusing, in a special group of CG-CNVs, patients with sSMC where the same correlation with infertility may be observed, it was shown that more than 50% of these cases inherited their CG-CNV from one of their parents [Manvelyan et al., 2008]. Maybe the recently suggested two-hit model [Girirajan et al., 2010] for CNV larger than 500 kb in size can resolve this snag in the future.

3.2. DE NOVO CG-CNVs

Although for MG-CNVs identified by aCGH new variants can be detected in each single generation [Girirajan et al., 2011], a de novo heterochromatic CG-CNV is only extremely rarely substantiated [Craig-Holmes et al., 1975; Verma et al., 1993] (Figure 6). Euchromatic CG-CNVs seem to behave like MG-CNVs, and their de novo formation has already been proven [Barber, Zhang et al., 2006].

Considering the sheer size of the heterochromatic blocks in 1q12, 9q12, 16q11.2, Yq12, and in the short arms of the acrocentric chromosomes, length variations in subsequent generations are to be considered, but may be below the resolution level of cytogenetics. Still, there are no studies available on that interesting point. Also, all these large heterochromatic blocks are not included in, or even only analyzable by, aCGH or NGS approaches.

According to the now known intergeneration variations in MG-CNV [Girirajan et al., 2011], variations in CG-CNVs are more than likely.

3.2.1. CG-CNVs and Tumors

From the 1960s to 1990s, CG-CNVs, among others, were suspected to be associated with an enhanced cancer risk. For example, there were discussions about a possible role of heteromorphic inversions of chromosomes 1, 9, and 16 in solid and/or lymphoid tumors, as well as in leukemia. Even though some studies showed surprisingly high concordance rates of the malignancy and special CG-CNV and/or heteromorphic inversions [Adhvaryu and Rawal, 1991] there were always other studies that could not substantiate those specific findings [Aguilar et al., 1981].

There are also rare reports of acquired CG-CNVs and heteromorphic inversions of chromosome 9 in cancer [González García et al., 1997; Eichler et al., 1997; Wan et al., 2000; Neglia et al., 2003; Betz et al., 2005; Udayakumar et al., 2009]. Those are considered as exotic findings without any meaning for the tumor evolution and progression. Still, they may reflect the general genetic instability of tumor cells [Bakhoum and Compton, 2012], as do amplifications with involvement of centromeric regions [Liehr et al., 1999; Sirvent et al., 2000; Neglia et al., 2003]. The finding of satellite III and α-satellite DNA expression under cellular stress (see Chapter 5, section 5.1.2.9) support, according to Hall et al. (2012), "the theory that if heterochromatic epigenetic marks are altered to a transcriptionally active state, the resulting overexpression of satellite sequences can lead to genomic instability and oncogenesis." As antagonist of this satellite expression the tumor suppressor Prep1 (also known as PKNOX1 or PBX/knotted 1 homeobox1 in humans) is implicated in controlling DNA damage and regulating histone methylation levels [Hall et al., 2012]. Also heterochromatic bands like 1q12 were reported to be involved in tumor-associated chromosomal rearrangements [Morerio et al., 2006; Millington et al., 2008].

On the other hand, specific MG-CNVs were already clearly linked to cancer susceptibility [Tchatchou and Burwinkel 2008; Kuiper et al., 2010]. Also, not surprisingly due to known general genetic instability of tumor cells [Bakhoum and Compton, 2012], de novo MG-CNVs have been detected exclusively in cancer but not in normal tissue of the diseased individual [Walker et al., 2012].

Overall, CG-CNVs are at present not considered meaningful for any kind of tumor predisposition or prognosis; for MG-CNVs there is also no such implication.

3.3. MG-CNVs

MG-CNVs can either be inherited or caused by de novo mutations of different sizes. Several molecular mechanisms are known to be responsible for the occurrence of submicroscopic CNV within the genome [Gu et al., 2008] (see Chapter 4, section 4.8).

Formation of CG-CNVs

Studies and literature on the formation of CG-CNVs are sparse. Nonetheless, in this chapter (possible) modes of CG-CNV formation are briefly discussed. For MG-CNV, see section 4.8.

4.1. ACROCENTRIC CHROMOSOMES' SHORT-ARM VARIANTS

CG-CNVs of the short arms of the acrocentric chromosomes can be reduced or increased in size. As mentioned in Chapter 2, section 2.1, the normal/average length of acrocentric short arms equals the length of chromosome 18p of the same metaphase (Figure 4A). In addition to length, the structure of an acrocentric short arm may be altered, leading to the emergence of a CG-CNV (see Chapter 5, section 5.1.1).

Since parts of other chromosomes have been shown to be involved in acrocentric short-arm enlargement and/or structural changes (see Chapter 5), it can be suggested that one way to alter the chromosome arm size/structure is meiotic translocation due to unequal crossing over [Ferguson-Smith, 1974; Farrell et al., 1993]. Translocations may lead to derivatives like der(15) t(15;22)(p11.2;p13); that is, 15p− and der(22)t(15;22)(p11.2;p13), and 22p+, or der(13)t(Y;13)(q12;p11.2) = 13p+ (Figure 7; see also Chapter 5). As shown in Figure 7A, acrocentric p-arm material may also be translocated to the end of another chromosome, besides another acrocentric short arm (Figure 7B).

Translocation of Yq12 heterochromatin to the short arm of an acrocentric chromosome is nonrandom: chromosome 15 is most often involved (52%), followed by chromosomes 22 (33%), 21 (7%), 13 (4%), and 14 (4%). This distribution of Yq/acrocentric short-arm translocations may reflect a differing degree of homology between subsets of classical satellite DNAs,

with the closest homology being shared between 15p, 22p and Yq. Furthermore, during the prophase of male meiosis, the XY bivalent is frequently seen in close proximity to an NOR. (It has been) suggested that this may be the reason for the relative frequency with which Yq/acrocentric short arm translocations are observed. [Wilkinson and Crolla, 1993]

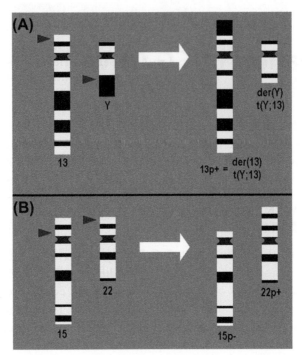

Figure 7 Acrocentric chromosomes' short-arm variants may be due to translocations. **A.** A reciprocal meiotic translocation may lead to one gamete with a 13p+ and another gamete with a derivative Y chromosome with acrocentric short-arm material at its proximal long arm. **B.** An unequal crossing over between the short arms of two acrocentric chromosomes, here chromosomes 15 and 22, leads to alterations in the short arm lengths, here a 15p– and a 22p+ variant. Arrowheads mark the chromosomal breakpoints.

Also rarely, acrocentric chromosomes' short-arm variants may be formed by a mitotic translocation event [Giussani et al., 1996; Storto et al., 1999].

Amplification or loss of short-arm material during meiosis cannot be neglected either. Still, it has already been shown that if different regions and DNA stretches were amplified they induce this kind of CG–CNV. Gain or reduction of acrocentric short-arm length may be caused by NOR (Figure 1) or satellite DNA stretches; also DNA derived from another acrocentric p-arm may be inserted or translocated. These changes can be characterized only by molecular cytogenetics using the corresponding probes (see Chapter 5). Mechanisms for this kind of acrocentric-p-arm amplifications are not clear but may be similar to those satellite-DNA amplifications seen in the centromeric regions (see section 4.2).

4.2. VARIANTS OF THE CENTROMERIC REGIONS

Heteromorphisms for all centromeric regions can be found in clinical diagnostics using GTG-banding and/or FISH. C-banding size differences have been found at relatively high frequencies as well. However, the latter have practically never been investigated systematically [Lubs et al., 1977].

Centromeric CG-CNVs are reported as loss or gain of alphoid DNA copy numbers (Figure 8). All these so-called cen− (Figure 8A) or cen+ (Figure 8B) variants have the same shortcoming: the only benchmark to which a stronger or weaker signal intensity of a FISH signal most often is made reference is the homologous second chromosome with normal signal intensity. Standardization could be achieved by cohybridization with a single copy probe, to which the signal intensities of centromeres in question could be compared [Liehr T, unpublished data]. Also, computer-assisted measurements were suggested for signal quantification [Iourov et al., 2005].

Rarely, not only is there amplification of α-satellite DNA but also duplication of the centromere itself, described for example for chromosome 9 [Kosyakova et al., 2013] or chromosome 11 [Till et al., 1991] (Figure 8C). Because the pericentromeric regions (also called centromeric transition regions) contain many segmental duplications [Bailey et al., 2001; Horvath

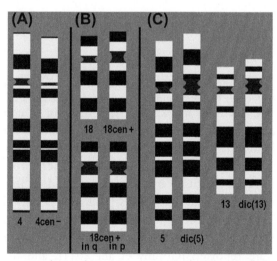

Figure 8 Centromere polymorphisms are summarized schematically here. **A.** reduction in size for a centromere of chromosome 4 (4cen-). **B.** amplification of a centromere of chromosome 18 (18cen+); may also be only in proximal p or proximal q. **C.** pseudo-dicentric (dic) chromosomes exemplified for a chromosome 5 and a chromosome 13.

et al., 2001 and 2003; She et al., 2004], those regions could be causative for the formation of those rare pseudodicentrics.

The amplification of certain DNA stretches, including centromeric DNA as well as dicentric chromosomes, is thought to happen by the so-called breakage fusion bridge (BFB) mechanism [McClintock, 1984; Moore et al., 2000]. This model

> proposes that the initiating event in amplification formation is the fusion of two chromosomes or chromatids, resulting in a dicentric chromosome and the initial novel joint. When the two centromeres segregate from each other during cell division, the DNA bridge between the two centromeres breaks. The broken chromosome end is unstable and after replication, the two sister chromatids with unstable ends often fuse, producing a new novel joint and a second (possibly) dicentric chromosome that begins another cycle. (...) Subsequent BFB cycles can then increase the number of amplified copies (...) and the number of novel joints. [Moore et al., 2000]

The frequency of such events in centromeric regions is not known; for euchromatic regions a rate of $< 1 \times 10^{-9}$ amplifications per cell and generation has been estimated [Moore et al., 2000].

Heteromorphic patterns of centromeric DNA are also discussed to

> arise frequently within an alphoid array as the result of unequal crossover or random mutational events (i.e.) amplifications (...). Alternatively, unequal crossing-over on a larger scale would disperse portions of clusters to different regions of the alphoid array and might account for the differences in total length of the alphoid array in chromosomes from different individuals. [Ge et al., 1992]

Also combinations of unequal crossing over [Mashkova et al., 1998], so-called "saltatory amplification steps" and deletion and/or unequal crossing over are discussed to explain the formation of cen+ and cen− variants [Marçais et al., 1991]. As the variation within the alphoid array of the Y chromosome (which evolves along haploid chromosome lineages) is approximately equal to what is seen for other chromosomes, the main mechanism of formation variants may be based on intrachromatide or unequal sister chromatide exchanges [Willard, 1991]. The influence of the centromeric transition regions [She et al., 2004] on CG-CNVs of the centromeric regions remains to be elucidated. Rarely de novo cen+ variants are reported [Goumy et al., 2011], and there are hints that recombination in the centromeric regions is suppressed compared to the surrounding euchromatic regions at least 10-fold [Laurent et al., 2003].

Finally, at least one case with a mitotic origin of a centromeric CG-CNV was described in a derivative chromosome 22 in a malignancy [González García et al., 1997].

4.3. VARIANTS OF NONCENTROMERIC HETEROCHROMATIN

Variations in size of noncentromeric heterochromatin (i.e., CG-CNVs) account specifically for the regions 1q12, 9q12, 16q11.2, and Yq12. In 2001, it was suggested that segmental duplications within the human genome are causative for many kinds of rearrangements; the pericentromeric regions especially have an enhanced rate of them, including the heterochromatic regions and blocks [Bailey et al., 2001; Horvath et al., 2001 and 2003; She et al., 2004]. The previously mentioned BFB mechanism (section 4.2) could also account in part for such variants.

Unequal crossing over in meiosis can be causative for some of the size variations of autosomal chromosomes (Figure 9). Ferguson-Smith (1974) reports a case of de novo 16q11.2 amplification most likely due to unequal crossing over. An acquired reduction in size of 1q12 was reported in one case associated with frontotemporal dementia/schizophrenia; however, in this report it was not clear if more than band 1q12 was deleted [Gourzis et al., 2012]. Finally, Neglia et al. (2007) report an amplification of the 1q12 region in a colon cancer cell line. However, additional material was involved there and an unequal sister chromatid exchange is thought to be responsible for formation (Chapter 3).

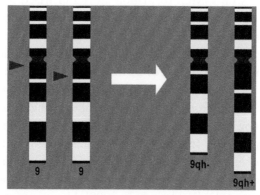

Figure 9 Unequal crossing over between two chromosomes 9 may lead to a 9qh+ and a 9qh− variant in meiosis. Arrowheads mark the chromosomal breakpoints.

Specific pericentric inversions of some chromosomes (e.g., 1, 3, or 9) are also considered heteromorphic variants. Even though CG-CNVs are not considered in this connection at first guess, there are hints that amplification of (peri)centromeric heterochromatic DNA is a promotor of this kind of inversion formation. In 1978 there already were first hints of this possibility [Verma, Dosik, Lubs, 1978], and later studies pointed toward the same direction [Cavalli et al., 1985, Starke et al., 2002]. Thus, pericentric inversions are also included in Chapter 5, section 5.1.2.

4.4. UNBALANCED CHROMOSOME ABNORMALITIES (UBCAs)

UBCAs without clinical consequences are present in more than one individual or within one family over more than one generation. Thus, in most cases the generation in which the UBCA evolved is not available for studies [Barber, 2005]. However, it is suggested that a UBCA may form due to unequal crossing over based on repeat units present in the genome at short distance, similar to that shown in Figure 9 for chromosome 9 heteromorphisms. Also, according to literature, unequal crossing over followed by a two-break reciprocal translocation or a three-break insertion could be the reason for this kind of CG-CNV formation [Jalal and Ketterling, 2004].

4.5. SMALL SUPERNUMERARY MARKER CHROMOSOMES (sSMCs)

Facts and theories on sSMCs formation are summarized in Liehr (2012). Overall, it is common sense that an sSMC is formed by the combination of one or more rare events happening during gameto- or embryogenesis. sSMCs may have different shapes, such as inverted duplicated, centric minute, and ring (Figure 10A). For formation of inverted duplicate-shaped sSMCs, inter- or intrachromosomal U-type exchange of homologous chromosomes resulting from a crossover mistake during meiosis is assumed (Figure 10B). For centric minute-shaped sSMC formation trisomic and monosomic rescue, post-fertilization errors and gamete complementation have been proposed in literature (Figure 10C). Finally, ring-shaped sSMCs may evolve by different, in part complex mechanisms, like the McClintock mechanism (Figure 10D) [Baldwin et al. 2008], U-type reunion, or reunion of the ends of centric minute-shaped sSMCs [Liehr, 2012]. Repetitive elements including fragile sites may be involved in their formation [Liehr, 2012].

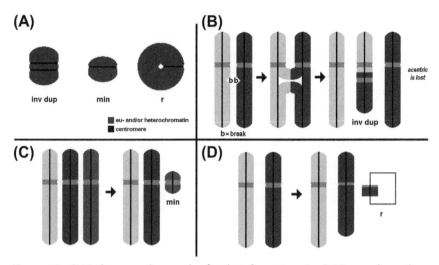

Figure 10 sSMC shapes and examples for their formation. **A.** sSMCs may have three shapes: inverted duplicated (inv dup), centric minute (min), and ring (r). **B.** Inverted duplication-shaped sSMCs may form due to crossing over and U-type exchange in meiosis. **C.** Centric minute-shaped sSMCs can form in connection with trisomic rescue as postzygotic event. **D.** Ring-shaped sSMCs can form due to the McClintock-mechanism shown here.

4.6. EUCHROMATIC VARIANTS (EVs)

EVs may form due to genomic flux and subsequent amplification of pseudogene cassettes [Barber et al., 1999]. But like EV regions, stretches with (hemi)heterochromatic features also are present, and formation may be based on similar mechanisms as discussed for centromeric (section 4.2) and heterochromatic CG-CNVs (sections 4.2 and 4.3).

4.7. GONOSOMAL-DERIVED CHROMATIN

Formation of variations in gonosomal-derived heterochromatin should mainly be due to loss of gain of copy numbers as described in sections 4.2 and 4.3. Gain or loss of gonosomal euchromatic material is either due to nondisjunction mechanisms [Jacobs, 1992] or translocations. Their formation was treated in sections 4.1 and 4.4.

4.8. MG-CNVs

Apart from ideas on amplification and loss of repetitive DNA stretches already mentioned in sections 4.2, 4.3, and 4.6, there is only one recent study that provided evidence that MG-CNVs may be amplified involving circular intermediates [Durkin et al., 2012]. In general, the major mechanisms for MG-CNV formation are considered to be nonallelic homologous recombination (NAHR) for recurrent rearrangements and nonhomologous end joining (NHEJ) for nonrecurrent rearrangements. NAHR can either be based on region-specific, low-copy-repeat (LCRs or segmental duplications), or sometimes repetitive sequences (e.g., Alu or LINE) as homologous recombination substrates [Lee et al., 2007]. When LCRs are located on the same chromosome and in direct orientation, NAHR results in deletion and/or duplication. Inversions result when LCRs on the same chromosome are in opposite orientation, whereas NAHR between LCRs located in different chromosomes result in translocation [Colnaghi et al., 2011]. Similar but more complicated are two recently reported mechanisms for triggering of (sub)chromosomal rearrangements: fork stalling and template switching (FoSTeS) and microhomology-mediated, break-induced replication (MMBIR) [Lee et al., 2007].

CHAPTER 5

Types of CG-CNVs

Here all CG-CNV variants published to date or characterized in my laboratory are summarized. I am happy to receive reports on further unpublished CG-CNVs or literature not included here unintentionally, to be incorporated in the next edition of this book (contact Thomas.Liehr@ med.uni-jena.de).

5.1. HETEROCHROMATIC CG-CNVs

This chapter on heterochromatic CG-CNVs includes acrocentric short arm variants (Chapter 2; section 5.1.1), the pericentric regions of all chromosomes (Chapter 2; section 5.1.2), and the autosomal noncentromeric heterochromatin (Chapter 2; section 5.1.2), including partly the EV of chromosomes 9 and 16 (Chapter 2; section 5.1.2), as well as the Y-chromosome-derived heterochromatin (Chapter 2; section 5.1.3).

5.1.1. Acrocentric Chromosomes' Short Arms

Deviations from the typical length of an acrocentric short arm are considered heteromorphisms, or CG-CNVs. The first ever such reported deviations were p-arm enlargements in patients with Marfan syndrome in 1960 [Tjio et al., 1960]. As for other CG-CNVs, loss or gain of length in acrocentric short arms was initially suggested to always be something meaningful in terms of clinical diagnostics. It was suggested, for example, that the p-arm length of acrocentrics might be correlated with Down syndrome frequency [Dekaban et al., 1963], which later was disproved [Sands, 1969]. However, there is still the problem that in some instances an acrocentric p+ or p− chromosome is indicative for a derivative chromosome with clinical impact (section 5.1.1.3). Also, an acrocentric short arm may have a regular length but be structurally abnormal in GTG, CBG, NOR, or fluorescence staining. Such cases can be due to amplification and simultaneous shortening of other parts of acrocentric short arm material [Conte et al., 1997] or translocation of other chromosomal material; such variants are also included in sections 5.1.1.1.1, 5.1.1.1.2, and 5.1.1.2.

Three components can be enlarged or lost (section 5.1.1.2) in acrocentric short arms:

- Satellite I, II, III (also classical satellite), and ß-satellite DNA in band p11.2
- Nucleolus organizing region (NOR); that is, multiple copies of ribosomal RNA-genes (genes coding for 18S and 28S rRNA and ribosomal proteins that coalesce to form the nucleolus) and ß-satellite-DNA in band p12
- ß-satellite-DNA, satellite I, and telomeric sequences present in p13

Furthermore, it was shown that satellite I DNA present on chromosomes 3q11.2 and 4q11 also covers the heterochromatic short arms of the acrocentric chromosomes [Tagarro et al., 1994; Fernández et al., 2001]. However, there also seem to be homologous sequences to the acrocentric short arms located in 5pter, as an early subtelomeric probe showed (BAC RP11-114j18) [Knight and Flint, 2000].

Band p11.2 constitutes the proximal acrocentric short arm, band p12 the satellite stalk, and p13 the satellite [Gardner and Sutherland, 2012] (Figure 11A). Unfortunately, the identical designation "satellite" is used for the cytogenetic band p13 and for α-, ß-satellite and satellite I to III-DNAs. In the first case satellite refers to the structure of band p13, which almost seems like it is not attached to the acrocentric chromosomes short arm ends, since band p12 often is barely visible in GTG-banding. In band p12 there is an average of about 40 copies of the NOR gene, making in summary 300 to 400 copies per cell [Wellauer and Dawid, 1979].

The name satellite in connection with DNA stretches has another origin; that is, in studies on human DNA, performed by cesium-chloride- or cesium-sulfate-based isopycnic centrifugation in the 1970s [Lee et al., 1997]. Using this kind of approach, besides a main peak there were also a few side peaks detectable, which were called satellite peaks containing satellite DNA [Manuelidis, 1978]. Later it turned out that main parts of these satellite DNAs are localized in the centromeric regions, including 1q12, 9q12, 16q11.2, Yq12, and the acrocentric short arms.

The first detected centromeric satellite family, between 1967 and 1971, were the classical satellite DNAs: satellite I (alternating 17 and 25 bp repeats), satellite II (alternating 5 bp repeats ATTCC and ATTCG), and satellite III (5 bp repeats ATTCC occasionally interspersed by A-TorG-TCGGGTTG). α-satellite DNA (171 bp repeats in human) was originally isolated from the African green monkey genome in 1971. ß-satellite DNA (68–69 bp repeats) was first described in 1981; γ-satellite DNAs (220 bp basic repeats in human)

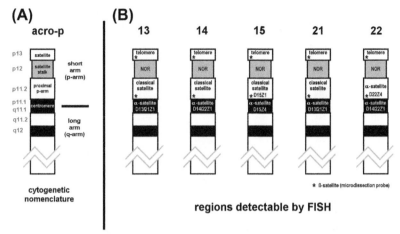

Figure 11 Schematic depictions of the short arms of the acrocentric chromosomes: **A.** Scheme of any acrocentric chromosomes with short and long arm designation (right side, black); cytogenetic nomenclature of the bands (left side, gray), and designations of the three bands of the short arm. Be aware that band p13 is called "satellite", which is not to be mixed up with satellite-DNA mentioned in part B of this figure. **B.** Localization of different regions of satellite DNA, of the NOR and the telomeres in the five different acrocentric human chromosomes. These regions were selected according to the availability of FISH probes used in diagnostics. Note that even though according genome browsers there are multiple BACs assigned exclusively to 21p, FISH studies revealed [Liehr, unpublished data] that there are homologous sequences on all other acrocentric short arms.

were reported between 1993 and 1995. A not-specially denominated 724 sequence family was first seen in 1984, and a 48 bp repeat family described in 1988 [Lee et al., 1997] later was called centromeric repeats (CER) [Warburton et al., 2008].

In Figure 11B for each human acrocentric chromosome the normal constitution of the short arm is depicted [Gardner and Sutherland, 2012], with special attention to the currently available FISH probes. Apart from that, there are, in parts, more detailed molecular analyses for the structure of the acrocentric short arms available in literature [Trowell et al., 1993; Müllenbach et al., 1996; Shiels et al., 1997; Bandyopadhyay et al., 2001; Piccini et al., 2001].

5.1.1.1. Enlargement of Acrocentric Chromosomes' Short Arms
The average size of an acrocentric short arm corresponds to the length of 18p of the same metaphase spread [Verma, Gogineni et al., 1977]. Enlargements should be considered as p+ if the short arm is larger than the length of 17p of

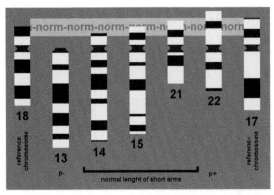

Figure 12 The normal size of an acrocentric short arm is between half of 18p and up to the full length of 17p. If the p-arm of an acrocentric chromosome is longer than 17p it is a p+ variant; if it is shorter than half the length of 18p it is a p– variant. Accordingly chromosome 13 in this depiction is a p– variant; chromosome 22 a p+ variant; and chromosomes 14, 15, and 21 have normal-sized short arms.

the same metaphase spread (Figure 12). In rare cases, acrocentric p–arms may be tremendously enlarged: they may become even longer than their actual euchromatic long arm [Lau et al., 1979; Friedrich et al., 1996].

5.1.1.1.1. Amplification of acrocentric short arm material

Enlarged short arms of one or more of the acrocentric chromosomes possibly also with acrocentric short arm reduction (see section 5.1.1.2), are regularly detected but normally not further analyzed in molecular detail. According to Table 1, acrocentric p+ variants are found in about 3% of the cases. Many of those are studied further in routine cytogenetics by CBG, NOR, and/or inverted DAPI staining (see also Figure 1), sometimes even by FISH approaches (see later), but rarely reported in literature (Table 6) [Chen et al., 1981]. The most frequently observed variants are schematically depicted in Figure 13 and may be distinguished by pure cytogenetic, nonmolecular approaches. However, using them it cannot be determined if an acrocentric p-arm enlargement is due to amplification (Table 6) or translocation of acrocentric short arm material (section 5.1.1.1.2). Sometimes even both mechanisms can be involved [Friedrich et al., 1996]. However, only in one family with enlargement of an already enlarged short arm was one chromosome 15 observed from one generation to the next [Manzanal Martínez et al., 1992].

Amplification of acrocentric p13 material can be described according to ISCN 2013 as an s+ variant, for enlarged satellite; p12-material is amplified in the case of an stk+ variant, for enlarged satellite stalk; an add(acro)(p13)

Table 6 Reported Amplifications of Acrocentric p-arm Material

p-arm subband chromosome	p11.2		p12		p13	
	cytogenetic	molecular	cytogenetic	molecular	cytogenetic	molecular
13	Soudek, 1979	n.a.	Bajnóczky and Meggyessy, 1985	Gar'kavtsev et al., 1986	Liehr, unpublished data	n.a.
14	n.a.	Choo et al., 1992	Liehr, unpublished data	Tantravahi et al., 1981; Cheng et al. 1989	Matsuda et al., 1989	Earle et al., 1989
15	Babu et al., 1986	Acar et al., 1999; Bucksch et al., 2012	Werner and Herrmann, 1984; Velázquez et al., 1991	Friedrich et al., 1996; Bucksch et al., 2012	Zhuang et al., 1994	Starke et al., 2005
21	n.a.	n.a.	Liehr, unpublished data	Fu, 1989; Bucksch et al., 2012	Murthy et al., 1990	n.a.
22	Mandal et al., 2003; Conte et al., 1997	Bernstein et al., 1981	Liehr, unpublished data	Liu et al., 1993; Langer et al., 2001; Bucksch et al., 2012	Liehr, unpublished data	n.a.

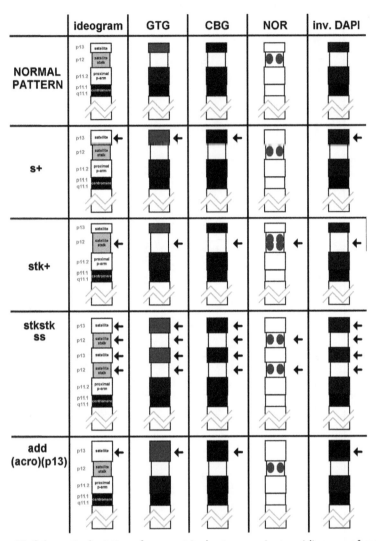

Figure 13 Schematic depiction of acrocentric short arm variants as idiogram, after GTG, CBG, NOR, and inverted DAPI (inv. DAPI) staining. The normal variant is shown in the first line. The variants s+, stk+, stkstk/ss, and add(acro)(p13) can be distinguished using the four staining approaches. The CG-CNV regions of interest are highlighted by arrows; in NOR staining not necessarily all changes are visible.

variant can be indicative for translocation of Yq12 material to an acrocentric chromosome (section 5.1.1.1.2.2). Amplification of band p11.2 cannot be further characterized by cytogenetic approaches because this region is hardly distinguished from the centromeric band p11.1-q11(.1). However for chromosomes 15 and 22 corresponding FISH probes are available.

Amplification of specific p-arm bands as reported in literature are summarized in Table 6. For amplification of band p12 it was shown that such amplified NOR regions tend to be inactive [Tantravahi et al., 1981].

Note: Amplification of any of the p-arm subbands may just lead to a structurally altered acrocentric short arm but not to a real p+ variant.

The CG-CNV stkstk/ss (Figure 13) is also typically visible as an enlarged acrocentric p-arm in banding cytogenetics. Most likely, it must be considered as duplication and/or resulting from a translocation of one acrocentric p-arm to another (Figure 7B; section 5.1.1.1.2).

ISCN 2013 suggests a nomenclature for heterochromatic variants of acrocentric short arms as summarized in Figure 13. Surprisingly, a PubMed [http://www.ncbi.nlm.nih.gov/pubmed] search for chromosome & stk, chromosome & stk+, chromosome & ss, chromosome & s+ or chromosome & ps+ gives almost no results with respect to chromosomal heteromorphisms. The latter may indicate that this nomenclature is not well accepted, like the recent invention of a new abbreviation psat+ for enlarged acrocentric short arms showed [De la Fuente-Cortés et al., 2009].

According to my own molecular cytogenetic studies [Liehr, unpublished data] on aberrant acrocentric short arms at least the following p+ variants have been detected so far:

1. Enlargements due to amplification of band p11.2 can only be verified for chromosomes 15 and 22 using the satellite III DNA probe D15Z1 (Color plate 1, Figure I) [Liehr et al., 2003] or the α-satellite-probe D22Z4 (Color plate 1, Figure II). In other words, currently existing enlargements of 13p11.2, 14p11.2, and 21p11.2 cannot be studied in more detail due to lack of specific FISH probes. According to Tagarro et al. (1994) satellite I DNA is present only in 13p11.2 and not in p11.2 bands of the other acrocentric chromosomes. This probe thus could be used to test 13p11.2 heteromorphisms, but it has not been done yet. *It is suggested that these aberrations be described as der(15)(p112amp) or der(22)(p11.2amp).*

2. Gain of copy numbers and/or amplification of band p12, also including larger or smaller parts of p11.2 and p13, can easily be detected using a NOR-specific DNA probe. The following types can be seen in all acrocentric chromosomes:

   a. Two NOR signals may be due to dup(acro)(p11.2p12), dup(acro)(p12p13), or dup(acro)(p11.2p13). Direct or inverted duplications are theoretically possible. As visible in Color plate 1, Figure III, different variants exist.
   Describe this aberration as der(acro)(p11.2-p13x2).

    b. Three or four NOR signals, due to duplication of p12 and neighboring regions, also have been observed (Color plate 1, Figure IV).
    Describe these aberrations as der(acro)(p11.2-p13x3) and der(acro)-(p11.2-p13x4).

    c. Finally, amplification of NOR can appear, leading to a complete or almost complete coverage of the enlarged short arm with NOR sequences (Color plate 1, Figure V).
    Describe these aberrations as der(acro)(p12amp).

3. Amplification of satellite DNA present along p11.2 and p13 is also possible, even though they are not easily detectable due to lack of corresponding specific DNA probes; here the following forms can be observed:

    a. Satellite DNA amplification derived from p13; that is, with a NOR signal in the more proximal region (Color plate 1, Figure VI).
    Describe this aberration as der(acro)(p13amp).

    b. Satellite DNA amplification derived from p11.2; that is, with a NOR signal in the more distal region (Color plate 1, Figures I and II).
    Describe this aberration as der(acro)(p11.2amp).

    c. According to Tagarro et al. (1994) satellite I DNA is present in all p13-bands of the acrocentric chromosomes; however, this probe has not been applied for CG-CNV-testing yet.

4. Also acrocentric variants may rarely be due to paracentric p-arm inversions like inv(acro)(p13p12) (Color plate 1, Figure VI) or complex inverted duplications (Color plate 1, Figure VII).
Note: Such CG-CNVs may easily be mixed up with small paracentric inversions, which can possibly have clinical impact in the offspring of the carrier; such a case was recently seen in my lab as inv(14)(p12q11.2).

NOR or CBG staining and a probe for all acrocentric short arms should be used to identify such acrocentric p-arm variants. Also helpful may be probes for satellite DNAs D15Z1 and D22Z4, and/or a FISH probe for the NOR region.

5.1.1.1.2. Addition of heterochromatic material to the acrocentric short arms

In addition to amplification of specific parts of acrocentric chromosome short arms (section 5.1.1.1.1), entire acrocentric short arms may be added as one or more copies to a normal acrocentric p-arm (section 5.1.1.1.2.1), or a nonacrocentric chromosome derived heterochromatin may be added to an acrocentric short arm (section 5.1.1.1.2.2).

5.1.1.1.2.1. Unbalanced translocations between acrocentric short arms: Enlargement of acrocentric short arms were observed and described as stkstk/ss in Figures 1 and 13 for all five human acrocentric chromosomes [Bauchinger and Schmid, 1970; Watson and Scrimgeour, 1977; Miller et al., 1978; Trabalza et al., 1978; Murthy et al., 1989; Percy et al., 1993; Conte et al., 1997; Bucksch et al., 2012; Liehr, unpublished data]. Different breakpoints seem to be possible in those variants, even leading to dicentric chromosomes in healthy carriers in rare cases (Figure 14) [Hancke and Miller, 1985; Ramos et al., 2008] (Color plate 1, Figure VIII). (*Describe this aberration as ss.*)

Such variants may also be cryptic, such as the case of a derivative chromosome 22 with centromeric material derived from chromosome 13 or 21 (D13/21Z1) and chromosome 22 (D14/22Z1) [Cockwell et al., 2003], which were not cytogenetically visible as ss-variants. Translocation of a short arm of chromosome 15 to another acrocentric chromosome is most often observed. This is visualized by probe D15Z1 in 15p11.2. D15Z1 is frequently present in the (peri)centromere of a chromosome 14 ($\sim$12% of the population) and is inherited within families [Smeets et al., 1992; Stergianou et al., 1993; Shim et al., 2003; Cockwell et al., 2007]. D15Z1 can also be present on chromosome 13 ($\sim$4% of the population) [Smeets et al.,

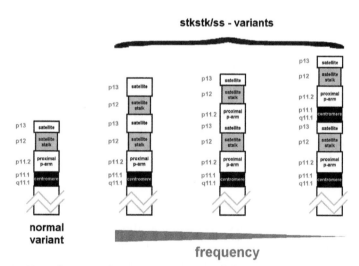

Figure 14 Normal variant of an acrocentric p-arm (left sided) and CG-CNV of stkstk/ss resulting from different breakpoints are schematically shown. The triangle below the stkstk/ss variants indicates the frequency in which those variants can be expected [Liehr, unpublished data].

1992; Cockwell et al., 2007], chromosome 21 ($\sim$1% of the population) [Wilkinson and Crolla, 1993; Cockwell et al., 2007], or chromosome 22 ($\sim$0.5% of the population) [Cockwell et al., 2007] (Color plate 2, Figure II). Since D15Z1 was not always present where expected in these derivative chromosomes but in exceptional cases in acrocentric short arm bands p13 or p12, more complex rearrangements involving insertions and inversions can be suspected. Also, due to the high frequency of this kind of CG-CNV in about 1% of the population the additional signal of D15Z1 was present on two acrocentric chromosomes simultaneously [Cockwell et al., 2007].

All these ss variants may be due to either meiotic unequal crossing over (Figure 7B) [Smeets et al., 1991; Peterson et al., 1992] or intrachromosomal duplication. Meiotic unequal crossing over was demonstrated to be causative for acrocentric short arm material derived from paternal 21p moving to fetal chromosome 13p [Farrell et al., 1993]. However, mitotic events leading to mosaicism have been observed as well [Livingston et al., 1985; Farrell et al., 1993; Guissani et al., 1996; Levy et al., 2000].

Also variants with three or four short arms have been observed as exceptional derivatives of acrocentric chromosomes in healthy carriers. Those were derived from one (Color plate 2, Figure I) or different acrocentric chromosomes (Color plate 2, Figure II) [Reddy and Sulcova, 1998]. In some cases the origin was not clarified either [Spowart, 1978; Pérez-Castillo et al., 1986].

A complete replacement of an acrocentric short by another is very rarely observed. It may result from two centromere-near breaks in the q(!)-arms of the involved chromosomes [Cockwell et al., 2003].

NOR or CBG staining and a probe for all acrocentric short arms should be used to identify such acrocentric p-arm variants. Also helpful may be probes for satellite DNAs D15Z1 and D22Z4, and/ or for the NOR region (see also probe recommendation at the end of the next section).

5.1.1.1.2.2. Unbalanced translocations involving acrocentric short arms and other heterochromatic material: The short arms of chromosomes 13, 14, 15, 21, and 22 may be enlarged due to unbalanced translocations/insertions involving only heterochromatic and geneless material of other chromosomes as well. This kind of imbalance may include several megabasepairs in size. Even though all heterochromatic regions could be involved, up to now exclusively Yq12 material has been observed

here. To the best of our knowledge neither 1q12, 9q12, nor 16q11.2 has ever been reported to contribute to an enlarged acrocentric short arm.

Rearranged chromosomes like a der(13)t(Y;13)(q12;p11.2) may be formed (see Figure 7A). Comparable chromosomes have been reported for all five acrocentric chromosomes—der(13)t(Y;13) [Bajnóczky et al., 1985; Morris et al., 1987; Doneda et al., 1992; Bucksch et al., 2012], der(14) t(Y;14) [Buys et al., 1979; Ellis et al., 1990; Bucksch et al., 2012], der(15) t(Y;15) [Verma, Dosik, Jhaveri, Worman, 1978, Spowart, 1979; Alitalo et al., 1988; Neumann et al., 1992; Yoshida et al., 1997; Rajcan-Separovic et al., 2001; Zhao et al., 2004; Meza-Espinoza et al., 2006; Chen et al., 2007; Bucksch et al., 2012; Onrat et al., 2012; Chen-Shtoyerman et al. 2012], der(21)t(Y;21) [Burk et al., 1983; Fernández et al., 1994; Cockwell et al., 2003], and der(22)t(Y;22) [Burk et al., 1983; Morales et al., 2007].

According to Smith et al. (1979) among the translocations involving Yq heterochromatin and the short arm of an acrocentric chromosome, chromosome 15 is the one most frequently observed (52%), followed by chromosomes 22 (33%), 21 (7%), 13 (4%), and 14 (4%) (see Chapter 4, section 4.1). However, different breakpoints can be present in different der(acro)t(Y;acro) (Color plate 2, Figure III) [Wilkinson and Crolla, 1993].

Although the aforementioned der(acro)t(Y;acro) chromosomes were found in healthy carriers and transmitted through generations, mitotic origin is also possible in mosaic cases like 45,X,der(14)t(Y;14)/46,XY [Andersson et al., 1988]. Furthermore, a de novo der(21)t(Y;21) of meiotic origin has already been reported [Ng et al., 2006]. In all de novo cases the derivative chromosomes need to be studied in detail by molecular approaches.

Most surprisingly there are several reports of add(acro)(p13)-CG-CNVs, which did not stain by Yq12 specific probes. Such variants were also found for all five acrocentric chromosomes. They catch one's eyes by an intense DAPI staining [Spowart, 1979; Burk et al., 1983; Haaf et al., 1989; Fernández et al., 1994, Verma, Kleyman, and Conte, 1996; Serakinci et al., 2001]. Serakinci et al. (2001) showed that the detectable telomeric repeats were not terminal in the corresponding enlarged chromosome 22p, but interstitial; that is, the amplification of material and insertion of amplified material was seemingly distal from the telomeres of this derivative chromosome.

More than 20 similar cases were seen in my laboratory [Liehr, unpublished data] (Color plate 2, Figure IV). Apart from the idea that this material could be an expansion from some low repetitive AT-rich DNA [Serakinci et al., 2001] or (GACA)n simple repeats [Schmid et al., 1994], we could also think of insertion and amplification of mitochondrial DNA. Interestingly, there are also two ssMCs reported that do not stain by any DNA probe at all [Liehr et al., 2008].

> **DAPI-staining and a probe for Yq12 should be used to identify and subclassify those two harmless variants (see also probe recommendation at the end of section 5.1.1.1.2.1). In case a de novo origin of the derivative acrocentric chromosome cannot be excluded, more detailed molecular analyses are recommended, for such cases as where a balanced translocation was observed in one of the parents.**

5.1.1.1.3. Addition of acrocentrics short arms to nonacrocentric q-arms or p-arms

The results of the most frequently observed translocation of short arm acrocentric material to another chromosome is the der(Y)t(Y;acro)(q12;p11.2) (Figure 7B; Color plate 2, Figure Va-1). This kind of derivative Y-chromosome with addition of an acrocentric p-arm part is stable over generations [Genest, 1972] without adverse effects on male fertility. Interestingly, here the pseudoautosomal region of the Y-chromosome in Yqter is lost [Kühl et al., 2001]. This der(Y) seems to be the reciprocal product of the event leading to the der(acro)t(Y;acro) discussed in section 5.1.1.1.2.2. This Y-chromosome variant, also called satellited Yq chromosome, was first reported in 1967 [Genest et al., 1967] and was later repeatedly studied using different (molecular) cytogenetic approaches [Genest, 1972; Howell et al., 1978; Shabtai et al., 1981; Giraldo et al., 1981; Bayless-Underwood et al., 1983; Schmid et al., 1984; Martin Lucas et al., 1984; Tsita et al., 1989; Wilkinson and Crolla, 1993; Couturier-Turpin et al., 1994; Verma, Gogineni et al., 1997; Penna Videaú et al., 2001; Bucksch et al., 2012]. Similar derivatives of Y-chromosome and an acrocentric short arm can be found as a de novo event in Turner syndrome cases [Hoshi et al., 1998]; even one Turner syndrome with a t(X;acro) [Stetten et al., 1986] and another one with a der(7) t(Y;7)(p11.1 ~ 11.2;p22.3) is reported [Polityko et al., 2009]. There is also one report on a phenotypically normal male with satellited short arm of the Y-chromosome [Lin et al., 1995].

Acrocentric short arms can be translocated to the ends of all human chromosomes. In most de novo cases this goes together with an adverse phenotypic effect [Faivre et al., 1999] and/or complex mechanisms of formation [Kleczkowska et al., 1988; Ki et al., 2001]. As the topic of this book are CG-CNVs no review is given on the satellited derivatives with adverse or unknown outcome but this may be found elsewhere [Sarri et al., 2011].

Still, there are reports on derivatives transmitted through generations for 1pter [Habibian et al., 1994; Bucksch et al., 2012], 1qter [Park et al., 1992], 2qter [Bauld and Ellis, 1984; Elliot and Barnes, 1992; Lamb et al., 1995; Reddy and Sulcova, 1998], 3qter [Liehr, unpublished data], 4pter [Estabrooks et al., 1992; Arn et al., 1995], 4qter [Babu et al., 1987; Arn et al., 1995; Miller et al., 1995; Shah et al., 1997; Guttenbach et al., 1999; Zaslav et al., 2004] (Color plate 2, Figure Vb), 10qter [Storto et al., 1999], 12pter [Willatt et al., 2001], 15qter [Liehr, unpublished data],17pter [Soudek, 1973; Verma et al, 1979; Patil and Bent, 1980; Killos et al., 1997], 18qter (Color plate 2, Figure Vc), 20pter [Park and Rawnsley, 1996], 20qter [Pimentel et al., 1989], 21qter (Color plate 2, Figure Vd), 22qter (Color plate 2, Figure Ve). In these latter cases no relevant material was lost on the chromosome to which the acrocentric derived short arm was translocated. In the case reported by Storto et al. (1999) the corresponding derivative chromosome originated mitotically in the father of the prenatally detected case. Only in one case α-satellite DNA was detected together with the translocated acrocentric short arm [Shah et al., 1997]. One case was observed where the derivative chromosome 3 carrying an acrocentric short arm was involved in another balanced translocation; here the long arm with the acrocentric p-arm was translocated to another acrocentric long arm, and a derivative acrocentric with two acrocentric p-arms was formed [Liehr, unpublished data].

However, in case one parent has a corresponding balanced translocation the offspring is likely to suffer from clinical problems induced by imbalances due to the presence of only one of the two derivative chromosomes [Chen et al., 2000; Sarri et al., 2011]. Rarely, carriers of large paracentric inversions within an acrocentric chromosome q-arm may produce offspring with a dicentric chromosome [Whiteford et al., 2000] (Figure 15; Color plate 2, Figure VI).

Parental cytogenetic analysis and the corresponding subtelomeric FISH probe should be applied to distinguish harmful from harmless variants, also helpful may be a probe for all acrocentric short arms. In case of a der(4)t(4;acro) probe D14/22 might be applied [Shah et al., 1997].

5.1.1.1.4. Insertion of acrocentric short arms to other chromosomes
A highly unexpected kind of heteromorphic variant or CG-CNV is the insertion of acrocentric short arm-derived chromosomal material into another chromosome. Such instances are rare but may have to be considered

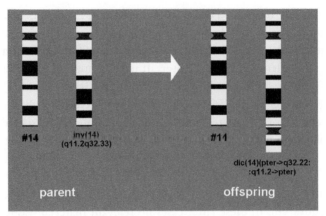

Figure 15 If a parent has a large paracentric inversion within an acrocentric chromosome q-arm (here chromosome 14), inversion loop formation may produce an offspring with a dicentric chromosome. In most cases (not necessarily all) this will lead to substantial imbalances.

in cases of an altered GTG banding result in a phenotypically normal person and a euchromatic variant is unlikely (Chapter 2, section 2.6 and section 5.2). Such harmless insertions have been reported for 6q15 [Prieto et al., 1989], 7p13 [Reddy and Sulcova, 1998], 7q21.3 ~ 22.1 [Guttenbach et al., 1998], 8q11 [Guttenbach et al., 1999], 11q21 [Cosper et al., 1985], 12p11 [Watt et al., 1984], 22q [Norris et al., 1995], 16q11.2 [Bucksch et al., 2012] (Color plate 2, Figure VII), and Yq [Norris et al., 1995]. The coinsertion of centromeric material (here derived from chromosome 21) was reported for only one case [Watt et al., 1984]. Even a de novo formation was observed for insertion of 14p12 material into 6p22; here the causative mechanism was an inversion loop formation due to a paternal balanced reciprocal translocation involving the satellite stalk region of an acrocentric chromosome [Chen et al., 2004].

However, if such kind of insertion damages a gene it may be disease-causing [Verellen-Dumoulin et al., 1994] or found in clinical cases as a causative finding or coincidental [Tamagaki et al., 2000; Chen et al., 2004; Bucksch et al., 2012]. Also insertion of p-arm material may appear in tumor, a topic not included in this book; see an example in Atkin and Baker (1995).

> **Parental cytogenetic analysis and a FISH probe for all acrocentric short arms should be applied to identify such variants. However, it is hard to distinguish harmful from harmless variants.**

5.1.1.2. Loss of Material of Acrocentric Chromosomes' Short Arms

As depicted in Figure 12 an acrocentric variant p– should be considered only if the short arm is less than half the size of chromosome 18p of the same metaphase spread. According to Table 1 such variants are approximately 20 times less frequently observed than p+ variants, but are inherited as other CG-CNVs in a Mendelian fashion [De los Cobos et al., 1981]. They were observed for chromosomes 13 [Emerit et al., 1968; Gebauer et al., 1988; Liehr et al., 2003], 14 [Emerit et al., 1972; Nielsen et al., 1978; Earle et al., 1992; Choo et al., 1992], 15 [Caglayan and Gumus; 2010], 21 [De los Cobos et al., 1981; Gaál et al., 1984; el-Badramany et al., 1989; Kar et al., 1992; Conte et al., 1996; Vorsanova et al., 2002; Alkhalaf et al., 2002], and 22 [Benítez et al., 1979; Izakovic, 1984; Conte et al., 1997]. Molecular studies are scarce, however loss of satellite III DNA (subband p11.2) [Earle et al., 1992; Choo et al., 1992], NOR region [Miller et al., 1977], or both [Conte et al., 1997] were demonstrated in exceptional cases. Among all acrocentrics most frequently loss of 21p material is reported; this derivative chromosome even got its own name in early cytogenetics: Christchurch (Ch1) chromosome, an aberration originally thought to be associated with chronic lymphocytic leukemia [Vorsanova et al., 2002].

According to my own molecular cytogenetic studies on aberrant acrocentric short arms [Liehr, unpublished data] at least the following p– variants are possible and distinguishable:

- Diminishing due to loss of (parts of) band p11.2 can only be verified for chromosomes 15 and 22 using the satellite III DNA probes D15Z1 and the α-satellite probe D22Z4 (Color plate 2, Figure VIII). For 13p11.2, 14p11.2, and 21p11.2 partial or complete loss cannot be studied at present or only indirectly by a small distance between centromere and NOR. *Describe this aberration as der(acro)(p11.2dim) or der(acro)(p11.2p11.2del).*
- Loss of copy numbers of band p12 can easily be detected using a NOR-specific DNA probe (Color plate 2, Figure IX). *Describe this aberration as del(acro)(p12p12).*
- For loss of ß-satellite DNA present along p11.2 and p13 the following forms can be observed:
  - Complete loss of the whole short arm (Color plate 3, Figure I) *Describe this aberration as del(acro)(p11.2) or del(acro)(p10).*

- Partial loss of p13 is possible (Color plate 3, Figure II) *Describe this aberration as del(acro)(p13p13)*.
- Other forms should also exist, but have not been reported yet.
- Finally, shortening of an acrocentric short arm may also be due to a small pericentric inversion as in a case seen in my laboratory with a final karyotype of 46,XX,der(21)(:q11.21->p11.1::q11.21->qter); however, here possibly the offspring can be affected, due to inversion loop formation, leading to unbalanced karyotypes.

NOR or CBG staining and a probe for all acrocentric short arms should be used to identify such variants. Also helpful may be probes for satellite DNAs D15Z1 and D22Z4, and/or the NOR region.

5.1.1.3. Altered Acrocentric p-arms as Hints on Cryptic Unbalanced Aberrations

It is the nightmare for every cytogeneticist to misinterpret an altered acrocentric short arm as a heteromorphism, which in reality is a meaningful chromosomal rearrangement. Thus, it is necessary (1) to report such potential chromosomal heteromorphisms, and (2) to study the parents. This statement is especially valid in prenatal diagnostics.

Reported cases with (half-)cryptic changes of acrocentric short arms connected with adverse clinical outcome are summarized in Table 7 (Color plate 3, Figure III). These cases can be due to a parental balanced rearrangement, which in many cases can more easily be recognized than the harmful derivative alone in the index case. Such derivatives can be de novo, and also in those cases a parental chromosomal study is helpful, as it may indicate that the suggested acrocentric short arm CG-CNV is nonexistent. According to Cockwell et al. (2003) five of 100 patients with mental retardation have cryptic aberrations in the acrocentric p-arms. Most of them are unbalanced translocations—one insertion was also reported [Vekemans et al., 1990]. According to Table 7 terminal parts of all human chromosomes may be involved; due to low case numbers it cannot be deduced if any chromosome is involved more often than the others.

Also p+ variants may appear in tumor, a topic not discussed in this book; see an example in Jenkyn et al. (1987).

Parental analysis, NOR, or CBG staining and probe for all acrocentric short arms as well as corresponding subtelomeric probes should be used to distinguish variants from meaningful rearrangements.

Table 7 Reported Addition of Euchromatin to Acrocentric p-arm Material

Addition of	To 13p	To 14p	To 15p	To 21p	To 22p
1			q: Villa et al., 2000; Verschuuren–Bemelmans et al., 1995		q: Chia et al., 1988
2	p: Al-Saffar et al., 2000				
3			q: De Marchi et al., 1977		q: Schwanitz et al., 1977
4		p: Gouw et al., 1972*		p: Furbetta et al., 1975	p: Forabosco et al., 1976; Schwanitz and Grosse, 1973; Metz et al., 1973
5	p: Dev et al., 1979		p: de Carvalho et al., 2008 p: Zhao et al., 1995Color plate 3, Figure IIIa		
6	p: Starke et al., 2005;		p: Engelen et al., 2001	q: Taysi et al., 1983	p: Hengstschläger et al., 2005
7		q: Bartsch et al., 1990			
8	p: Guanti et al., 1976			q: Ozdemir et al., 2012	

(Continued)

Table 7 Reported Addition of Euchromatin to Acrocentric p-arm Material—cont'd

Addition of	To 13p	To 14p	To 15p	To 21p	To 22p
9				q: Color plate 3, Figure IIIb	q: Sanger et al., 2005 p: de Chieri et al., 1978 q: Chen and Lin, 2003; Roux et al., 1974
10	p: Aller et al., 1979				
11	q: Smeets et al., 1997	p: Benzacken et al., 2001			
12				p: Hansteen et al., 1978	p: Allen et al., 1996; Dallapiccola et al., 1980
13	q: Vekemans et al., 1990			q: Di Bella et al., 2006	
14	q: Gilgenkrantz et al., 1990		q: Keyeux et al., 1990		q: Sutton et al., 2002
15	q: Annerén et al., 1982	q: Tatton-Brown et al., 2009	q: Tatton-Brown et al., 2009		
16	q: Savary et al., 1991				p: de Ravek et al., 2005

17		p: De Pater et al., 2000 q: Parcheta et al., 1985 q: Fryns, Logghe et al., 1979	p: Dufke et al., 2006 p: Spinner et al., 1993	q: Fryns, Parloir et al., 1979; Color plate 3, Figure IIIc p: Jacobsen and Mikkelsen, 1968
18				
19	p or q: Starke et al., 2005			
20		p: Lurie et al., 1985	p: Color plate 3, Figure IIId	p: LeChien et al., 1994
21				
22				q: Feenstra et al., 2006
X	p: Gustashaw et al., 1994	p: Schempp et al., 1985		
Y				q: Butomo et al., 1984 q: Boyd et al., 2005

*Could also be tc 15p.

5.1.2. Pericentric Regions of All Chromosomes

Here CG-CNVs of the bands p11(.1) to q11(.1) of all human chromosomes plus bands 1q12, 9q12, and 16q11.2 are covered. The first 24 regions are considered as centromeres necessary for the formation of the kinetochor during cell division; they consist of α-satellite tandem DNA repeats of 171 bp, which are not annotated in any of the human genome browsers under their names (e.g., D7Z1). Approximate average sizes are mentioned for each chromosome according to literature [Choo et al., 1991]. The bands 1q12, 9q12, and 16q11.2 are treated here as well, because heteromorphisms can often be detected only as alterations of the corresponding pericentric regions plus their heterochromatic blocks. Also, EVs of chromosomes 9 and 16 are included in this chapter, since they are parts of corresponding pericentric CG-CNVs. Overall, CG-CNVs are reflected in this section as gain or loss of copy numbers of pericentric and centromeric cytobands. Some frequent pericentric inversions are included as well, because they may go together with copy number changes or look similar to gain of heterochromatic DNA stretches.

Molecular studies have shown that all centromeric regions are heteromorph [Choo et al., 1991] and may contain single base changes even in exceptional cases detectable by FISH approaches [O'Keefe et al., 1997] and inherited in a Mendelian fashion [Alexandrov et al., 2001]. Not all these heteromorphisms in banding cytogenetics lead to visible CG-CNVs [Bonfatti et al., 1993], and only a minority of those (molecular) cytogenetically visible centromeric variants were reported in literature. Overall, these so-called cen+ and cen− variants are to be expected for each and every human centromere. In the following chromosome-specific subsections, this statement is in parts amended by my own observations of such variants [Liehr, unpublished data]. Finally, radioactive in situ hybridization approaches might be more suited to study and determine CG-CNV differences in centromeric regions than the now applied nonradioactive FISH approach [Yurov et al., 1987].

Note: For interphase FISH analysis: Although amplification of centromeric DNA is not a big deal, cen− variants with complete absence of α-satellite DNA may lead to suggest chromosomal monosomy of the sample studied, which is then a false-negative result [Sokolic et al., 1999; Duval et al., 2000; Tsuchiya et al., 2001]. Variants leading to additional copies of α-satellite DNA on other chromosomes may lead to suggest chromosomal trisomy of the sample studied, which is a false-positive result (see section 5.1.2.25).

All centromeric regions consist of repetitive DNA stretches. As mentioned before (Chapter 2, section 2.2 and section 5.1.1), there are different so-called satellite DNAs present, but also other short and long interspersed nuclear elements (SINEs and LINEs) [Prades et al., 1996]. Besides, α-satellite is present at every centromere, apart from neocentromeres, and consists of 171 bp repeat units. The α-satellite family may be grouped in three subfamilies characterized by sequence similarity due to evolutionary relatedness. Chromosomes 1, 3, 5, 6, 7, 10, 12, 16, and 19 are within subfamily 1; chromosomes 2, 4, 8, 9, 13, 14, 15, 18, 20, 21, and 22 are in subfamily 2; and chromosomes 1, 11, 17, and X are in subfamily 3 [Matera et al., 1993]. The α-satellite DNA of the Y-chromosome is differently structured; that is, it only has 170 bp repeat units [Cooper et al., 1993]. An α-satellite DNA block itself is highly homologous within a chromosome-specific stretch and forms so-called higher order repeat units (HORs) that correlate with centromere function. HORs are characterized by a distinct repeating linear arrangement of an integral set of 171 bp monomers [Warburton et al., 2008]. For the X-chromosome it was shown that at the edge of the centromeric HOR α-satellite block there are approximately 40 kb of related but not identical DXZ1 monomers with decreasing homology (down to about 70% only) to the original 171 bp stretch DXZ1 [Schuler et al., 2001].

At the centromere also present may be ß-satellite-DNA; satellite I, II, and III DNA (section 5.1.1); and γ-satellite DNA. The latter has been identified on chromosomes 8 (704 bp repeats), 12 (216 bp repeats), X (1,205 bp repeats) [Warburton et al., 2008], and Y (200 bp repeats) [Lee et al., 2000]. Other repeats were also recently reported, like hsat4 (35 bp repeats) [Schueler et al., 2001; She et al., 2004], hsatII (may have 23, 26, 49, or 98 bp repeat units) [Warburton et al., 2008], and many others [She et al., 2004]. Some other repeats were described in the 1990s and not further characterized [Lee et al., 1997]. Some of such DNA repeat units are listed as chromosome-specific, because they also might contribute to CG-CNVs even though they are not yet tested for that in most cases.

Finally, a recent finding was that the (peri)centromeres of all human chromosomes accumulate during evolution duplications and thus are enriched for those. Centromeric transitions of 4q11, 5p11, 6q11, 8p11, 16q11, 18q11, 19q11, and Xp11 are relatively sharp; that is, there are only small stretches of duplicons. In contrast the following regions show extensive zones of duplications: 1q11–1q12, 2p11, 2q11, 6p11, 7p11, 7q11, 9p11, 9q11, 10p11, 10q11, 11p11, 13q11, 15q11, 16p11, 17p11, 18p11,

21q11, 22q11, and Yp11. The remaining studied regions show an average level of duplications along the centromeric transition: 1p11, 3p11, 3q11, 4p11, 5q11, 8q11, 11q11, 12p11, 12q11, 14q11, 17q11, 19p11, 20p11, 20q11, Xp11, and Yq11. Many if not all of these transitions seem to accumulate sequences and duplicons homologous to and thus derived from other chromosomes [Horvath et al., 2000; Vermeesch et al., 2003; She et al., 2004; Grunau et al., 2006].

5.1.2.1. Chromosome 1

- Regular size of centromeric α-satellite DNA D1Z7 (formerly pC1.8, pZ5.1 or pE25a): ~1,900 kb [Carine et al., 1989; Choo et al., 1991; Finelli et al., 1996; Alexandrov et al., 2001]
- Cytobands: 1p11.1 to 1q11
- Centromere size according to GRCh37/hg19: 7,400 kb (see Table 4)
- Regular size of centromeric α-satellite DNA D1Z5 (formerly pAL.1 or pSD1-1): 440 to 1,900 kb [Willard and Waye, 1987; Finelli et al., 1996; Lee et al. 1997; Alexandrov et al., 2001];
- Cytobands: 1q11 to 1q12
- Size according to GRCh37/hg19: 13,700 kb (see Table 4)

The α-satellite DNA stretch D1Z7 is present on chromosome 1 as well as on chromosomes 5 and 19 in the respective centromeric regions; there they are denominated as D5Z2 and D19Z3, respectively. Size variants of α-satellite DNA are reported as 1cen− and 1cen+ (Color plate 3, Figure IVa) [Brito-Babapulle, 1981].

The α-satellite DNA D1Z5 is present along the cytoband 1q12; according to Finelli et al. (1996), D1Z5 is also detectable in 1p11.1. Also located in 1q12 is satellite I DNA [Fernández et al., 2001], the satellite II DNA D1Z1 [Warburton et al., 2008], and ß-satellite DNA [Meneveri et al., 1993]. Also three blocks of long terminal repeat (LTR) arrays were identified in 1q12 [Warburton et al., 2008].

In band 1q12, variants can appear as 1qh− [Schwanitz, 1976] and 1qh+ [Kim, 1975; Schwanitz, 1976] (Figure 1; Color plate 3, Figure V). The latter has an approximate frequency of 0.25% in the general population (Table 1), while 1qh− variants are extremely rare. As shown in Figure 4, a qh+ variant should be considered only if band 1q12 is larger than a chromosome 16p of the same metaphase spread. A 1qh− would be present if the band is smaller than half of 18p of the same metaphase spread. 1qh+ variants fall into two cytogenetic groups: those that are homogeneously stained and those that have one or more extrabands [Maegenis et al., 1978] (Color plate 3,

Figure V). The extrabands seem to be due to specific packing of the DNA in the enlarged band 1q12, however their origin is not really understood. Up to four such extrabands were observed in an enlarged 1q12 region [Liehr, unpublished data]. Similar findings are reported for 9q12 and Yq12 (see sections 5.1.2.9 and 5.1.3.1).

Furthermore, there are inversion heteromorphisms that may be caused by homologies in the regions 1p12 and 1q13 [Hardas et al., 1994; Weise et al. 2005] (Color plate 3, Figure VI). They are reported as inv(1)(p11q12) [Gardner and Sutherland, 2004] or inv(1)(p13q21) [Brothman et al., 2006]. As shown in Color plate 3, Figure IVb, the breakpoints are proximal of the BAC probes RP11-130B18 and RP11-35B4; that is, they are proximal of positions 115,919,903 in the p-arm and of 143,999,790 in the q-arm (hg18). As the regions involved in this inversion do not contain any genes, inversion loop formation (if existent at all) cannot cause problems here. Besides there is one extraordinary case with a similar inversion reported, where the derivative chromosome 1 seemed to be pseudodicentric; possibly here a pseudodicentric chromosome 1 (see Figure 8C) acquired an inversion heteromorphism [Verma, Ramesh et al., 1996].

Even though satellite DNA normally is not thought to be transcribed, satellite III RNA derived from 1q12 was detected in senescent cells and some cancer cells [Enukashvily et al., 2007] (see also section 5.1.2.9).

To characterize CG-CNVs of chromosome 1 centromeric region at least probes D1Z7 and/or D1Z5 should be applied. Possibly centromere-near/centromere-flanking probes in 1p12 and 1q21.1 could be used as well [Liehr et al., 2006] to identify pericentric inversions. Finally, partial chromosome paints may be helpful (Color plate 3, Figure VI).

5.1.2.2. Chromosome 2
- Regular size of centromeric α-satellite DNA D2Z1 (formerly pX2 or M81229): 1,050-2,900 kb [Haaf and Willard, 1992; Lo et al., 1999; Alexandrov et al., 2001]
- Cytobands: 2p11.1 to 2q11.1
- Centromere size according to GRCh37/hg19: 6,300 kb (see Table 4)

Evolutionarily the centromere of chromosome 2 should be one of the youngest primary constrictions in human, since chromosome 2 derived from a fusion of two ancestral chromosomes [Charlieu et al., 1993]. This fusion distinguishes *Homo sapiens* from all other great apes, having 48 chromosomes [Hamerton et al., 1963].

2cen+ (Color plate 3, Figure VIIa) and 2cen– variants can be found as CG-CNVs, as already shown in 1992 by means of molecular genetics [Haaf and Willard, 1992]; they can be visualized by the D2Z1-α-satellite probe. There is also another alphoid probe denominated pBS4D [Rocchi et al., 1990], which hybridizes to centromeric region of chromosome 2 and 20 simultaneously.

Band 2p11.2 was proven to harbor an at least 26 kb of satellite $CAGC_n$ DNA repeat and a hsatII repeat with a minimal size of 20 kb [Warburton et al., 2008]; also satellite II DNA was detected [Lee et al., 1997] together with LTR arrays in 2q11 [Warburton et al., 2008]. Furthermore, the region adjacent to D2Z1 was studied in detail recently [Horvath et al., 2001; She et al., 2004].

Like chromosome 1 (see section 5.1.2.1), chromosome 2 can exhibit a pericentric inversion, which is regarded as a heteromorphism [Brothman et al., 2006; Gilling et al., 2006; Fickelscher et al., 2007]. This inv(2)(p11.2q13) is as frequent as 0.11% in the general population (Table 1). Four different variants were identified in European population, one of them being responsible for the inversion in almost 90% of the cases. Overall for inv(2) heteromorphism it could be shown that

> the majority of cases were not related, demonstrating (...) that the inv(2)(p11.2q13) is a truly recurrent rearrangement. Therefore, (...) the majority (of inversions) have arisen independently in different ancestors, while a minority either have been transmitted from a common founder or have different breakpoints at the molecular cytogenetic level. [Fickelscher et al., 2007]

The inversion may be characterized by the 2p11.2 breakpoint-spanning BAC probe RP11-153P14 and the two 2q13 breakpoint-flanking BACs RP11-399B17 and RP11-438K19 (Color plate 3, Figure VIIb).

To characterize CG-CNVs of chromosome 2 centromeric region at least probe D2Z1 should be applied. To identify the most frequent inv(2) variant the following probes can be used: RP11-153P14 (2p12) and RP11-1429F20 together with RP11-80K12 (2q13).

5.1.2.3. Chromosome 3

- Regular size of centromeric α-satellite DNA D3Z1 (formerly p3-9, pHS05, pAEQ.68 or Z12006): 2,900 to 3,407 kb [Yurov et al., 1987; Waye and Willard, 1989; Choo et al., 1991; Lo et al., 1999; Alexandrov et al., 2001]
- Cytobands: 3p11 to 3q11.1
- Centromere size according to GRCh37/hg19: 6,000 kb (see Table 4)

Applying GTG banding chromosome 3 centromere does not seem to be very heteromorph or prone to be associated with CG-CNVs. In contrary, using DA/DAPI/fluorescence banding centromere 3 provides many different variants [Gardner and Sutherland, 2004]. The latter CG-CNVs can be broken down to even more varieties, which are based on pericentric satellite I DNA [Tagarro et al., 1994; Buño et al., 2001]. It was shown that satellite I DNA also present on some other chromosomes covers the heterochromatic band 3q11.2 [Tagarro et al., 1994].

Using FISH with D3Z1 as alphoid probe 3cen+ and 3cen− CG-CNVs were delineated [Yurov et al., 1987]. However, no relationship was observable between FISH results and the DA/DAPI/fluorescence banding patterns [Kruminia et al., 1988]. Also, a second alphoid sequence was mentioned in 1991 (called VIIB4 spanning at least 2,750 kb), which was not further characterized [Choo et al., 1991].

Molecular analysis can define more variants (point mutations) of α-satellite DNA D3Z1, which are inherited in a Mendelian fashion [Waye and Willard, 1989]. Also other approaches showed a so-called lateral asymmetry; that is, heteromorphisms of centromere 3 [Brito-Babapulle, 1981].

Finally, as mentioned for chromosome 1, (pseudo)dicentric variants of chromosome 3 were seen (Figure 8C; Color plate 4, Figure Ia).

The pericentric inversion inv(3)(p11.2 ~ 13q12) is one heteromorphic variant of chromosome 3 [Brothman et al., 2006; Gilling et al., 2006] (Color plate 4, Figure Ib); the latter was determined to have a frequency of about 2% in the general population [Soudek and Sroka, 1978; Ibraimov et al., 1986]. It was suggested that small alphoid DNA stretches in 3q12 may contribute to the formation of this kind of inversion [Conte et al., 1992]. For Finland another potentially harmless pericentric inversion was reported: inv(3)(p12q24) [Lindberg et al., 1992]. Also a single report on an inv(3)(p11q11) variant is available [Balkan et al., 1983].

To characterize CG-CNVs of chromosome 3 centromeric region at least probe D3Z1 should be applied. Possibly centromere-near/ centromere-flanking probes in 3p12 and 3q13 could be used as well [Liehr et al., 2006] to identify pericentric inversions.

5.1.2.4. Chromosome 4

- Regular size of centromeric α-satellite DNA D4Z1 (formerly pZ4.1 or Z12011): 3,200–5,000 kb [Mashkova et al., 1994; Alexandrov et al., 2001]
- Cytobands: 4p11 to 4q11
- Centromere size according to GRCh37/hg19: 4,500 kb (see Table 4)

Similar to what is known about the centromeric region of chromosome 3 (section 5.1.2.3), the corresponding domain in chromosome 4p11-q11 shows also different CG-CNVs in DA/DAPI/fluorescence banding [Harris et al., 1987; Gardner and Sutherland 2004]. Two main types of heteromorphism were observed: either (1) an intense fluorescence band in the centromeric region or (2) one in the proximal area of the short arm; both were seen in about 8% of the studied chromosomes 4 [Bardhan et al., 1981] (Figure 8B). In only a minority of the cases these variants can be substantiated by banding cytogenetics [Docherty and Bowser-Riley, 1984].

Satellite I DNA, which is also present on a few other chromosomes, covers the heterochromatic band 4q11 [Tagarro et al., 1994]. Also the satellite II DNA repeat $GAATG_n$ was assigned to 4p11 and 4q11, spanning at least 65 and about 27 kb, respectively. In addition, megasatellite arrays approximately 50 kb in size and two blocks of LTR arrays were found in 4p11 [Warburton et al., 2008]. These DNA stretches have not been applied in molecular cytogenetics yet for the characterization of chromosome 4 heteromorphisms.

Applying the alphoid DNA stretch D4Z1, 4cen− and 4cen+ CG-CNVs can be found [Brito-Babapulle, 1981; Liehr et al., 2003; Zaslav et al., 2004] (see section 5.1.2.25). Interestingly, some of those may give the impression of a pericentric inversion in banding cytogenetics [Liehr et al., 2003].

Also, there are two alphoid probes called p4n1/4 [D'Aiuto et al. 1997] and pG-Xba11/340 [Hulsebos et al., 1988] that hybridize to centromeres of chromosomes 4 and 9 and are present 2,000 to 4,000 times in the genome [Hulsebos et al., 1988].

Note: Under low-stringency conditions D4Z1 may produce cross-hybridization on centromere of chromosome 9 [Petit et al., 1998] and vice versa for D9Z4.

To characterize CG-CNVs of chromosome 4 centromeric region at least probe D4Z1 should be applied. Possibly centromere-near/centromere-flanking probes could be used as well [Liehr et al., 2006] to exclude small pericentric inversions.

5.1.2.5. Chromosome 5

- Regular size of centromeric α-satellite DNA D5Z12 (formerly M26920): n.d. [Puechberty et al., 1999; Alexandrov et al., 2001]
- Cytoband: 5p11

- Regular size of centromeric α-satellite DNA D5Z2 (formerly pC1.8, pZ5.1 or M26919): ~2,000–4,000 kb [Choo et al., 1991; Finelli et al., 1996; Puechberty et al., 1999; Lo et al., 1999; Alexandrov et al., 2001]
- Cytobands: 5p11 to 5q11.1
- Regular size of centromeric α-satellite DNA D5Z1 (formerly pG-A16 or M26920): ~40-250 kb [Choo et al., 1991; Finelli et al., 1996; Puechberty et al., 1999; Alexandrov et al., 2001]
- Cytoband: 5q11.1
- Centromere size according to GRCh37/hg19: 4,600 kb (see Table 4)

Potential CG-CNVs of chromosome 5 centromere can further be characterized using the α-satellite probe D5Z2, which also gives signals on chromosomes 1p11.1-q11 (section 5.1.2.1) and 19p11-q11 (section 5.1.2.19). 5cen+ and 5cen– variants have been seen [Liehr, unpublished data], but have not been reported in literature yet. Additionally, there are two reports on large C-positive bands in 5q11.1, which turned out to be insertions of 9q12 material (see section 5.1.2.25) [Fineman et al., 1989; Doneda et al., 1998].

Besides D5Z2 two further chromosome-5-specific α-satellite DNA stretches were defined in one study [Hulsebos et al, 1988; Puechberty et al., 1999]: the stretch D5Z12 is located in 5p11, D5Z2 from 5p11 to 5q11.1, and D5Z1 can be found in 5q11.1. D5Z1 cross-hybridizes to identical DNA sequences on chromosome 19, denominated there as D19Z2 (section 5.1.2.19), but not to chromosome 1.

The inversion inv(5)(p12~13q13) is discussed controversially in literature; most authors consider it to be nondeleterious for the individual and its offspring [Brothman et al., 2006; Gilling et al., 2006], but it is questioned by others [Simi et al., 1991].

> **To characterize CG-CNVs of chromosome 5 centromeric region at least probe D5Z1 or D5Z2 should be applied. Possibly centromere-near/centromere-flanking probes could be used as well [Liehr et al., 2006] to exclude small pericentric inversions.**

5.1.2.6. Chromosome 6
- Regular size of centromeric α-satellite DNA D6Z1 (formerly p308, pEDZ6 or AB005791): ~3,000 kb [Choo et al., 1991; Jabs and Carpenter, 1988; Lo et al., 1999; Alexandrov et al., 2001]
- Cytobands: 6p11.1 to 6q11.1
- Centromere size according to GRCh37/hg19: 4,600 kb (see Table 4)

6cen+ variants were repeatedly reported and verified by C-banding [Madan and Bruinsma, 1979; Friedrich, 1979] and FISH using the α-satellite probe

D6Z1 [Jabs and Carpenter, 1988; Lin et al., 1994; Caglayan et al., 2009]; also extremely strong amplification of the D6Z1 stretches were observed in q-arm [Lin et al., 1994] (Color plate 4, Figure II) or p-arm [Goumy et al., 2011] (Figure 8B). 6cen− variants are not reported in the literature, but also were observed (Color plate 4, Figure II).

For chromosome 6 one of the rare centromere duplications is reported [Callen et al., 1990], as mentioned in Chapter 4, section 4.2 (Figure 8C). Also an inv(6)(p11.2q12) was observed once, which might be considered as heteromorphism as well [Liehr, unpublished data].

> To characterize CG–CNVs of chromosome 6 centromeric region at least probe D6Z1 should be applied. Possibly centromere-near/centromere-flanking probes could be used as well [Liehr et al., 2006] to exclude small pericentric inversions.

5.1.2.7. Chromosome 7

- Regular size of centromeric α-satellite DNA D7Z1 (formerly p308, pZ7.6B or M16087): ∼ 1,600 to 3,810 kb [Choo et al., 1991; Wevrick and Willard 1991; Finelli et al., 1996; Fetni et al., 1997; Alexandrov et al., 2001]
- Cytoband: 7p11.1
- Regular size of centromeric α-satellite DNA D7Z2 (formerly pMG87 or p7AL) [Choo et al., 1991; Finelli et al., 1996]: 100 to 500 kb [Fetni et al., 1997] or ∼ 5,500 kb [Wevrick and Willard 1991]
- Cytoband: 7q11.1
- Centromere size according to GRCh37/hg19: 3,700 kb (see Table 4)

The short arm part of centromere on chromosome 7 (7p11.1) contains D7Z2 alphoid DNA, and the long arm part 7q11.1 is covered by D7Z1 stretches [Wevrick and Willard 1991; Fetni et al., 1997; Choo et al., 1991]; both regions are separated by approximately a 1 Mb stretch of unknown DNA sequences [Fetni et al., 1997]. Interestingly, this order is reported in an inverse way by Finelli et al. (1996), which might be a hint on a possible polymorphism. CG–CNVs of chromosome 7 centromere can be detected using the alphoid probes D7Z1 and D1Z2.

In literature only examples for 7cen− variants are present, which were found during leukemia diagnostics. The latter gave false-positive results for a monosomy 7 in interphase evaluation [Duval et al., 2000]. Sokolic et al. (1999) pointed out that this 7cen− variant may not even be visible in GTG banding. 7cen+ variants were also seen [Liehr, unpublished data]. More detailed information about pericentromere of chromosome 7 and two other small alphoid stretches D7Z5 and D7Z6 in the short arm adjacent to D7Z2 can be found in literature [de la Puente et al., 1998]. Finally, band 7p11.21 was proven to harbor two hsatII repeats of at least overall 40 kb in size [Warburton et al., 2008].

> To characterize CG-CNVs of chromosome 7 centromeric region at least probes D7Z1 and/or D7Z2 should be applied. Possibly centromere-near/centromere-flanking probes could be used as well [Liehr et al., 2006] to exclude small pericentric inversions.

5.1.2.8. Chromosome 8

- Regular size of centromeric α-satellite DNA D8Z2 (formerly pJM18, pZ8.4 or M64779): ~ 1,000 to 2,550 kb [Choo et al., 1991; Ge et al., 1992; Lo et al., 1999; Alexandrov et al., 2001]
- Cytobands: 8p11.1 to 8q11.1
- Centromere size according to GRCh37/hg19: 5,000 kb (see Table 4)

For chromosome 8-centromere 8cen+ and 8cen− variants can be characterized using the alphoid probe D8Z2 [Liehr, unpublished data]. Also, D8Z1 is commercially offered as an alphoid probe for chromosome 8. However, according to Nusbaum et al. (2006) such a DNA stretch does not exist; it seems D8Z1 is identical to D8Z2 (Dr. I. Alexandrov, personal communication). In 8q11.1 a γ-satellite DNA stretch of at least 137 kb was recently detected and an LTR array in 8p11.1 [Warburton et al., 2008].

> To characterize CG-CNVs of chromosome 8 centromeric region at least probe D8Z2 should be applied. Possibly centromere-near/centromere-flanking probes could be used as well [Liehr et al., 2006] to exclude small pericentric inversions.

5.1.2.9. Chromosome 9

- Regular size of centromeric α-satellite DNA D9Z4 (formerly pMR9A or M64320): 2,700 kb [Rocchi et al., 1991; Finelli et al., 1996; Lo et al., 1999; Alexandrov et al., 2001]
- Cytobands: 9p11.1 to 9q11
- Centromere size according to GRCh37/hg19: 3,400 kb (see Table 4)
- Regular size of centromeric satellite III DNA D9Z3: n.a. [Fernández et al., 2001]
- Cytoband: 9q12
- Size according to GRCh37/hg19: 15,200 kb (see Table 4)

The α-satellite DNA D9Z4 is located in cytobands 9p11.1 to 9q11. There are also two potentially identical alphoid probes (p4n1/4 [D'Aiuto et al. 1997] and pG-Xba11/340 [Hulsebos et al., 1988]) that hybridize to centromeres of chromosomes 4 and 9 (see section 5.1.2.4). Furthermore, satellite I and III DNAs are located in 9q12, the latter denominated D9Z3 (formerly D9Z1) [Fernández et al., 2001], and a ß-satellite DNA in the flanking heterochromatic bands 9p11.1 to 9q12 called D9Z5 covering approximately 2,500 kb [Greig and Willard, 1992; Meneveri et al., 1993; Starke et al. 2002]. Warburton et al. (2008) identified additionally a so-called large tandem array family, which covers 10 distinct regions on chromosome 9 including 9p13, 9p11.2, and 9q12. Warburton et al. also described three blocks of LTR arrays in 9q11.2 and two blocks in 9q12. On the molecular cytogenetic level the (peri)centric region of chromosome 9 (Color plate 4, Figure IIIa) appears similar to that of chromosome 1 (Color plate 3, Figure VI) but not chromosome 16 (Color plate 4, Figure IIIb).

As well as 9cen+ and 9cen– [Kosyakova et al., 2013], there are many other variants including the 9q12 region (see section 5.1.2.25) and even the adjacent so-called hemiheterochromatic bands 9p12 and 9q13 [Starke et al., 2002]. The latter can hardly be distinguished in GTG banding (Figures 1 and 16). Banding cytogenetics may characterize just 9qh+, 9qh–, inv(9)(p11q13), and 9ph+ variants.

A qh+ CG-CNV is considered if band 9q12 is larger than a chromosome 16p of the same metaphase spread (Figure 4). A 9qh– would be present if the band is smaller than half of 18p of the same metaphase spread. 9qh+ variants fall into two cytogenetic groups—those that are homogeneously stained (at least in C-banding) [Steffensen et al., 2009] and those that have extrabands [Verma at al., 1993]. These extrabands may be an indication of a duplication involving 9q13 material, and it may appear if only 9q12 material is amplified [Kosyakova et al., 2013] (see sections 5.1.2.1 and 5.1.3.1). Variant 9ph+ is also considered an EV (see section 5.2.1.9).

Inversion heteromorphisms summarized cytogenetically as inv(9)(p11q13) show an overall population frequency of 2.86% (Table 1). However, inv(9)(p11q13) has variant frequencies in different human populations: 0.73% in Caucasians, 3.57% in African, 2.42% in Hispanic, and 0.26% in Asian (study size 6,250 individuals). In contrast, no such differences for 9qh+ were found; it seemed to be absent only in Asian populations [Hsu et al., 1987] and otherwise present in 0.25% of the general population.

By FISH over 25 variants can be subdivided, which can also in parts be present on the same chromosome 9 [Kosyakova et al., 2013] (Figure 16).

For chromosome 9, as for chromosomes 3 (see section 5.1.2.3) and 6 (see section 5.1.2.6), one of the rare centromere duplications is reported [Kosyakova et al., 2013] (Figures 8C and 22).

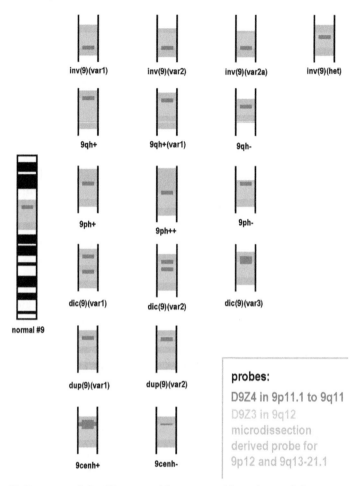

Figure 16 Fourteen of the 25 reported heteromorphic variants of chromosome 9 are depicted; also the most frequently observed normal state of the pericentric region of chromosome 9 using the probe set reported in Kosyakova et al. (2013) is presented (left). Variants of pericentric inversions (line 1), of bandsize q12 (line 2), of bandsize p12 (line 3), of dicentrics (line 4), of duplications (line 5), and centromeric size (line 6) are shown.

As reviewed by Madon et al. (2005), heat shock proteins associate after a corresponding stimulus to the heterochromatic subband 9q12. So called stress granules form primarily on 9q12 and also contain a

subset of RNA processing factors. Within these structures, heat shock transcription factor 1 (HSF1) binds to satellite III repeated elements and drives the RNA polymerase II-dependent transcription of these sequences into stable RNAs, which remain associated with the 9q12 region for a certain time after synthesis, even throughout mitosis. Other proteins, in particular, splicing factors, are also shown to relocalize to the granules upon stress. Evidence for transcriptional activity within a locus considered so far as heterochromatic and silent and the existence of a new major heat-induced transcript in human cells that may play a role in chromatin structure has also been revealed. [Madon, 2005]

In other words satellite III RNA is involved in the recruitment of RNA processing (i.e., splicing factors) during stress responses [Chen and Carmichael et al., 2010]. Besides transcription of satellite III DNA from 9q12 also α-satellite sequences of chromosome 12 and 15 were proven to be expressed after a heat shock [Rizzi et al., 2004] as well as HSF1 binding and transcription of satellite II and III sequences at pericentromeric regions of chromosomes 1, 5, 7, 9, 10, 16, and 17 as well as the short arms of all acrocentric chromosomes [Enukashvily et al., 2007; Eymery et al., 2010]. Also satellite III RNA expression is increased in progeroid laminopathies [Shumaker et al., 2006].

> **To characterize CG-CNVs of chromosome 9 centromeric region at least probes D9Z4 and D9Z3 or D9Z5 should be applied. Centromere-near/centromere-flanking probes could be used as well [Liehr et al., 2006; Kosyakova et al., 2013] to identify pericentric inversions and duplications. Finally partial chromosome paints may be helpful (Figure 16).**

5.1.2.10. Chromosome 10

- Regular size of centromeric α-satellite DNA D10Z1 (formerly pAlpha10-RP8,-RR6, pZ10-2.3 or M93288): 1,020 to 2,200 kb [Choo et al., 1991; Jackson et al., 1993; Lo et al., 1999; Alexandrov et al., 2001]
- Cytobands: 10p11.1 to 10q11.1
- Centromere size according to GRCh37/hg19: 4,300 kb (see Table 4)

The pericentric region of chromosome 10 was one of the first ones studied in detail. As expected, the α-satellite DNA D10Z1 can be heteromorph and 10cenh− [Devilee et al., 1988] as well as 10cenh+ CG-CNVs were observed [Liehr, unpublished data]. The alphoid DNA stretch is flanked by satellite III DNA (∼100–150 kb in proximal short arm and ∼320 kb in proximal long arm) [Jackson et al., 1993]. In addition, there are two hsatII repeats of at least 30 kb in length in band 10p11.1 [Warburton et al., 2008]. Satellite II DNA stretches of 78 to 530 kb and 28 to 620 kb were found in 10p11.1 and 10q11.1, respectively [Jackson et al., 1993, Warburton et al., 2008]. It was speculated that based on these sequences there is a non-deleterious inversion inv(10)(p11.2q21.2) for chromosome 10 [Brothman et al., 2006; Gilling et al., 2006] (Color plate 4, Figure IV). And indeed, satellite II-DNA stretches were found to be colocalized with the inversion prone regions 10p12.33 and 10q11.21.

To characterize CG-CNVs of chromosome 10 centromeric region at least probe D10Z1 should be applied. Possibly centromere-near/centromere-flanking probes could be used as well [Liehr et al., 2006] to exclude small pericentric inversions.

5.1.2.11. Chromosome 11

- Regular size of centromeric α-satellite DNA D11Z1 (formerly pCL11-A, pCL11-B, pHS53 or M21452): 850 to 4,760 kb [Yurov et al., 1987; Choo et al., 1991; Waye et al., 1987 Alexandrov et al., 2001; Lee et al., 1997]
- Cytobands: 11p11.1 to 11q11
- Centromere size according to GRCh37/hg19: 4,100 kb (see Table 4)

The centromere of chromosome 11 contains the alphoid sequence D11Z1, which can be present at normal signal size, as 11cenh+ or 11 cenh− CG-CNVs [Yurov et al., 1987]. Molecular genetically, these variations were already proven in 1987 [Waye et al., 1987], and a duplication of the centromere itself was reported twice [Till et al., 1991; Liehr, unpublished data].

To characterize CG-CNVs of chromosome 11 centromeric region at least probe D11Z1 should be applied. Possibly centromere-near/centromere-flanking probes could be used as well [Liehr et al., 2006] to exclude small pericentric inversions.

5.1.2.12. Chromosome 12

- Regular size of centromeric α-satellite DNA D12Z3 (formerly pB12, pAlpha12H8, pBR12 or M93287): 1,400 to 4,300 kb [Choo et al., 1991; Looijenga et al., 1992; Lo et al., 1999; Alexandrov et al., 2001; Lee et al. 1997]
- Cytobands: 12p11.1 to 12q11
- Centromere size according to GRCh37/hg19: 4,900 kb (see Table 4)

The centric and pericentric region of chromosome 12 was studied in detail by Vermeesch et al. (2003). Satellite I and II DNA were found in the proximal short arm, and another α-satellite DNA besides D12Z3 called s12A10 was detected in the proximal long arm of chromosome 12 [Vermeesch et al., 2003]. In 12p11.1 also a γ-satellite DNA was found to span at least 133 kb [Warburton et al., 2008].

12cenh– [Weier and Gray, 1992] and 12cenh+ variants [Chantot-Bastaraud et al., 2003; de Pater et al., 2006] were detected using the centromeric probe D12Z3. 12cen+ CG-CNVs can be seen as enlargement of the C-band positive region into the p-arm [de Pater et al., 2006] or the q-arm [Chantot-Bastaraud et al., 2003] (Figure 8B). Such variants may also cause the cytogeneticist to suggest an inversion as reported for other chromosomes [Liehr et al., 2003].

Transcription of α-satellite sequences of chromosome 12 was reported after induced heat shock [Rizzi et al., 2004].

To characterize CG-CNVs of chromosome 12 centromeric region at least probe D12Z3 should be applied. Possibly centromere-near/centromere-flanking probes could be used as well [Liehr et al., 2006] to exclude small pericentric inversions.

5.1.2.13. Chromosome 13

- Regular size of centromeric α-satellite DNA D13Z1 (formerly αRI or D29750): ∼1,800 to 2,300 kb [Trowell et al., 1993; Lo et al., 1999; Alexandrov et al., 2001]
- Cytobands: 13p11.1 to 13q11
- Centromere size according to GRCh37/hg19: 3,200 kb (see Table 4)

For the visualization of the centromere of chromosome 13 there is only one commercially available alphoid probe, D13Z1 (see section 5.1.2.25). It stains both, centromeres of chromosome 13 and 21; that is, D13Z1 is almost 100% identical to D21Z1 (see section 5.1.2.21). There were reports how to distinguish between chromosomes 13 and 21 by FISH [Soloviev et al., 1998] or primed in situ hybridization (PRINS) [Pellestor et al., 1994], however these were heteromorphic and identified only 70% of chromosomes 13 unambiguously [Yang et al., 2001].

The region covered by the DNA-stretch D13Z1 is one of the most polymorphic ones of the human centromeres. The prevalence of 13cen– CG-CNVs (here complete absence of signals after FISH) is 0.12% [Lo et al., 1999]. 13cen– also was repeatedly observed by others [Iurov et al., 1991; Pellestor et al., 1994; Nilsson et al., 1997] and 13cen+ CG-CNVs were also reported [Iurov et al., 1991; Nilsson et al., 1997; Liehr et al., 2003].

More detailed information about pericentromere of chromosome 13 and four other small alphoid stretches D13Z2, D13 Z3, D13Z4, and D13Z6 in the short arm adjacent to D13Z1 can be found in literature [Choo et al., 1991; Trowell et al., 1993]. Also satellite I DNA (pTRI-6) and satellite III

DNAs (D13Z5 and another denominated as pTRS-2) were located there [Lee et al., 1997]. In 13q11 an LTR array was identified [Warburton et al., 2008]. Interestingly, in all acrocentric centromeres including chromosome 13 there are sequences homologous to DNA located in 5qter, as an early subtelomeric probe showed (cosmid B22a4) [Knight and Flint, 2000]. Chromosome 13 centromeric CG-CNVs may also be due to insertion of D15Z1 material (see section 5.1.1.1.2.1).

> **To characterize CG-CNVs of chromosome 13 centromeric region at least probe D13Z1 should be applied, which also gives signals on chromosome 21. Possibly centromere-near/centromere-flanking probes could be used as well [Liehr et al., 2006] to exclude small pericentric inversions; also a short arm probe for all acrocentric chromosomes is helpful in many cases (section 5.1.1).**

5.1.2.14. Chromosome 14

- Regular size of centromeric α-satellite DNA D14Z1 (formerly αXT or D14Z9 or M22273): 1,360 to 2,300 kb [Trowell et al., 1993; Lo et al., 1999; Alexandrov et al., 2001]
- Cytobands: 14p11.1 to 14q11.1
- Centromere size according to GRCh37/hg19: 3,000 kb (see Table 4) 14cen+ [Dale et al., 1989] and 14cen– CG-CNVs can be characterized by the alphoid probe D14Z1, which is almost 100% identical to D22Z1 (see section 5.1.2.22). Even though a probe specific for 14p11.2 (pTR9-H2) was reported and an amplification of this region was detected by that probe, the latter was never in use in FISH-diagnostics [Earle et al., 1989].

More detailed information about pericentromere of chromosome 14 and three other small alphoid stretches D14Z4, D14Z6, and D14Z9 in the short arm adjacent to D14Z1 can be found in literature [Choo et al., 1991; Trowell et al., 1993]. Satellite I DNA (pTRI-6) and satellite III DNAs (D14Z7 and another denominated as pTRS-63) were also located there [Lee et al., 1997], as was a 123kb stretch of centromeric repeat-satellite DNA (CER) in 14q11.1 [Warburton et al., 2008]. Finally, in all acrocentric centromeres there are sequences that are homologous to stretches located in 5qter, as an early subtelomeric probe showed (cosmid B22a4) [Knight and Flint, 2000].

Also, variants on a chromosome 14 centromeric region may be due to insertion of D15Z1 material (see section 5.1.1.1.2.1).

To characterize CG-CNVs of chromosome 14 centromeric region at least probe D14Z1 should be applied, which also gives signals on chromosome 22. Possibly centromere-near/centromere-flanking probes could be used as well [Liehr et al., 2006] to exclude small pericentric inversions; also, a short arm probe for all acrocentric chromosomes is helpful in many cases (section 5.1.1).

5.1.2.15. Chromosome 15

- Regular size of centromeric α-satellite DNA D15Z4 (formerly pTRA-25): 2,500 to 4,500 kb [Choo et al., 1990; Choo et al., 1991; Finelli et al., 1996 and 2012]
- Cytoband: 15p11.1
- Regular size of centromeric α-satellite DNA D15Z3 (formerly pMC15 or AF237720): 2,500 to 4,500 kb [Choo et al., 1990; Finelli et al., 1996 and 2012; Alexandrov et al., 2001]
- Cytoband: 15q11.1
- Centromere size according to GRCh37/hg19: 4,900 kb (see Table 4)

To characterize region 15p11.1-q11.1 the alphoid stretches D15Z3 or D15Z4 may be used and 15cen+ and 15cen− CG-CNVs can be found. Choo et al. (1990) report another α-satellite stretch in the centromeric region of chromosome 15 called pTRA-20, for which heteromorphisms were proven as well [O'Keefe and Matera, 2000]. Also, satellite I DNA (pTRI-6) and satellite III DNA (pTR9-H2) were localized in pericentromere of chromosome 15 [Lee et al., 1997].

For cytoband 15p11.2 with a regular size of the satellite III DNA region of approximately 2,500 to 3,000 kb [Higgins et al., 1985; Choo et al., 1990] probe D15Z1 is available, which may also vary in size [Bucksch et al., 2012]. D15Z1 cen+ heteromorphism was described by Riordan and Dawson (1998) and are treated in section 5.1.1.1.2.1 (Color plate 1, Figure I). Variants of this region were proven as lateral asymmetry earlier as well [Brito-Babapulle, 1981]. Also, in all acrocentric centromeres there are sequences that are homologous to some located in 5qter, as an early subtelomeric probe showed (cosmid B22a4) [Knight and Flint, 2000].

Interestingly, transcription of α-satellite sequences of chromosome 15 was reported after heat shock [Rizzi et al., 2004].

To characterize CG-CNVs of chromosome 15 centromeric region at least probes D15Z3 or D15Z4 and D15Z1 should be applied. Possibly centromere-near/centromere-flanking probes could be used as well [Liehr et al., 2006] to exclude small pericentric inversions; also a short arm probe for all acrocentric chromosomes is helpful in many cases (section 5.1.1).

5.1.2.16. Chromosome 16

- Regular size of centromeric α-satellite DNA D16Z2 (formerly pSE16, pZ16A-2 or M58446): ~ 1,400 to 2,000 kb [Greig et al., 1989; Choo et al., 1991; Lo et al., 1999; Alexandrov et al., 2001]
- Cytobands: 16p11.1 to 16q11.1
- Centromere size according to GRCh37/hg19: 4,000 kb (see Table 4)
- Regular size of satellite II DNA D16Z3 (formerly pHuR195): n.a. [Savelyeva et al., 1994]
- Cytoband: 16p11.2
- Size according to GRCh37/hg19: 8,400 kb (see Table 4)

The α-satellite DNA D16Z2 can be used to characterize the chromosomal bands 16p11.1-q11.1. Size variants or CG-CNVs of α-satellite DNA are reported as 16cen− and 16cen+ [Liehr, unpublished data]. In literature, sometimes a probe D16Z1 is mentioned, which seems to be identical to D16Z2 [Myokai et al., 1999].

The satellite II DNA stretch D16Z3 is present in the large heterochromatic block 16q11.2 [Fernández et al., 2001]. Band 16p11.2 was proven to harbor three hsatII repeats of at least overall 100 kb in size, as well as satellite I, and III DNA in 16q11.2 [Fernández et al., 2001] and band 16p11.1 carries hsat4 units of at least 50 kb [Warburton et al., 2008]. A detailed map of centromeric transition in proximal 16p was published in the year 2000 [Horvath et al., 2000].

For band 16q11.2 also CG-CNVs were described as 16qh+ [Doyle, 1976; Thompson et al., 1990; Bucksch et al., 2012, case 3] and 16qh− [Hsu et al., 1987; Verma et al., 1992; Colls et al., 2004; Chatzimeletiou et al., 2006]. A 16qh+ variant is defined as 16q11.2 being larger than 16p; in 16qh− CG-CNVs 16q11.2 is smaller than 18p of the same metaphase spread (Figure 4). 16qh+ variants appear with a frequency of 0.37% of the general population (Table 1), while 16qh− are much less frequent (i.e., 0.016%) [Hsu et al., 1987].

Interestingly, for chromosome 16p11.2 no 16qh+ CG-CNV with additional band has been reported [Savelyeva et al., 1994], which is in contrast to observations in 1q12 and 9q12 as well as Yq12. One reason could be that the pericentric region of chromosome 16 is on the molecular cytogenetic level different than those of chromosome 1 (Color plate 3, Figure VI) and 9 (Color plate 4, Figure III); that is, it behaves like all other chromosomes when hybridizing partial paints specific for the corresponding chromosome arms. Still, at least in cases having a CG-CNV leading to a pseudodicentric chromosome der(16)(pter->q11.2::p11.1->qter), an extraband is visible [Liehr, unpublished data] (Color plate 4, Figure V).

> *Note:* If phenotypic abnormalities like developmental delay and mental retardation are observed with a potential 16qh+ variant, closer molecular cytogenetic studies of this region might be indicated as deleterious, and duplication of 16q12.2 to 16q13 might be involved [Barber, Zhang et al., 2006]. For 16ph+ variants see section 5.2.1.16.

A potentially harmless pericentric inversion is also reported for chromosome 16 as inv(16)(p11.2q12.1) [Brothman et al., 2006].

To characterize CG-CNVs of chromosome 16 centromeric region at least probes D16Z2 and D16Z3 should be applied. Possibly centromere-near/centromere-flanking probes could be used as well [Liehr et al., 2006] to exclude small pericentric inversions.

5.1.2.17. Chromosome 17

- Regular size of centromeric α-satellite DNA D17Z1-B: 500 to 900 kb [Rudd et al., 2006]
- Cytoband: 17p11.1
- Regular size of centromeric α-satellite DNA D17Z1 (formerly p17H8, TR17 or pYAM7-29 or M13882): 1,000 to 2,700 kb [Yurov et al., 1987; Choo et al., 1991; Lo et al., 1999; Alexandrov et al., 2001; Rudd et al., 2006]
- Cytobands: 17p11.1 to 17q11.1
- Centromere size according to GRCh37/hg19: 3,600 kb (see Table 4)

The D17Z1 stretch, most often used for centromere 17 detection, is accompanied by another, smaller alphoid DNA region in 17p11.1 called D17Z1-B [Rudd et al., 2006]. 17cen− [Lo et al., 1999] or 17cen+ CG-CNVs [Liehr, unpublished data] may be detected using centromeric probe D17Z1. The prevalence of 17cen− (here defined as complete absence of a signal) has been determined as 0.11% [Lo et al., 1999]. Alphoid single basepair variants were identified and already applied in molecular cytogenetic studies to distinguish homologous chromosomes from each other [O'Keefe et al., 1997].

To characterize CG-CNVs of chromosome 17 centromeric region at least probe D17Z1 should be applied. Possibly centromere-near/ centromere-flanking probes could be used as well [Liehr et al., 2006] to exclude small pericentric inversions.

5.1.2.18. Chromosome 18

- Regular size of centromeric α-satellite DNA D18Z2 (formerly pBS18AL or M65181): 1,700 kb [Finelli et al., 1996; Alexandrov et al., 2001]
- Cytoband: 18p11.1
- Regular size of centromeric α-satellite DNA D18Z1 (formerly 2XBA or M65181): 1,360 kb [Finelli et al., 1996; Devilee et al., 1986; Alexandrov et al., 2001]
- Cytobands: 18p11.1 to 18q11.1
- Centromere size according to GRCh37/hg19: 3,600 kb (see Table 4)

D18Z1 and D18Z2 belong to the alphoid subfamily II; they seem to be the result of two independent amplification events during evolution [Alexandrov et al., 1991]. Both regions may be lost or amplified as detectable, but not distinguishable by CBG-banding. Amplification may stretch into the q-arm [Beverstock et al., 1997] or p-arm; the latter 18cen+ variants are also sometimes designated as 18ph+ in literature [Verma and Babu, 1987; Pittalis et al., 1992; Bonfatti et al., 1993; Sensi et al., 1994] (for review see [Zelante et al., 1994]; Color plate 4, Figure VI). For the carrier of an 18ph+ reported by Sensi et al. (1994) the CBG-positive band was completely stained by a D18Z1 probe. In addition, a centromeric repeat-satellite DNA (CER) of at least 10 kb was located in 18q11.21 [Warburton et al., 2008].

The alphoid probe D18Z1 is one of the most frequently applied probes in prenatal interphase FISH diagnostics. Due to this a multitude of variants have been reported for this centromere (see also section 5.1.2.25), even though it is not considered as one of the chromosomes with a high prevalence of heteromorphisms [Lo et al., 1999]. In contrast to Lo et al., Verma et al. (1998) found over 40% of the chromosomes 18 have large-sized pericentric heterochromatin, whereas only 24% had medium-sized; the latter would be considered normal in other chromosomes.

18cen+ [Yurov et al., 1987; Verma and Babu, 1987; Quack et al., 1987; Sensi et al., 1994; Verma et al., 1998] and 18cen– CG-CNVs [Tepperberg et al., 2001; Thilaganathan et al., 2000; Weremowicz et al., 2001; Skinner et al., 2001; Bourthoumieu et al., 2010] were described. In some of those latter cases there were no detectable centromeric FISH signals on one chromosome 18p11.1–q11.1 [Bourthoumieu et al., 2010] (Figure 27B).

> **To characterize CG-CNVs of chromosome 18 centromeric region at least probe D18Z1 should be applied. Possibly centromere-near/ centromere-flanking probes could be used as well [Liehr et al., 2006] to exclude small pericentric inversions.**

5.1.2.19. Chromosome 19

- Regular size of centromeric α-satellite DNA D19Z2 (formerly pG-A16): n.d. [Choo et al., 1991; Hulsebos et al., 1988; Finelli et al., 1996]
- Cytoband: 19p11
- Regular size of centromeric α-satellite DNA D19Z3 (formerly pC1.8; PZ5.1 or M26919): ~2,000 kb [Choo et al., 1991; Finelli et al., 1996; Puechberty et al., 1999; Lo et al., 1999; Alexandrov et al., 2001]
- Cytoband: 19q11
- Centromere size according to GRCh37/hg19: 4,200 kb (see Table 4)

The centromere of chromosome 19 can be stained only by using probes D19Z2 or D19Z3, giving signals in chromosomes 5 and/or 1 as well (see sections 5.1.2.1 and 5.1.2.5). D19Z1 is located in 19p11, and D19Z3 in 19q11. 19cen+ and 19cen– CG-CNVs have been reported earlier based on CBG studies [Crossen, 1975; Gardner and Wood, 1979; Friedrich, 1985]. According to Crossen (1975) amplification of the CBG-positive region appears in 23% of all chromosomes 19 and can happen in proximal long arm (11%), both proximal arms (8%), or only the proximal short arm (4%). Also the very gene-poor regions 19p12 and 19q12 are not well studied; it remains unclear if they show variants as reported in earlier times of cytogenetics as 19qh+ [Doyle, 1976]. Still, band 19q12 was proven to harbor hsat4-repeats of at least 530 kb in size, and a variable number tandem repeat (VNTR) block of at least 50 kb in size was located in 19p12 [Warburton et al., 2008].

For 19cen+ CG-CNVs only amplification of D19Z3 has been observed so far (Color plate 4, Figure VII) [Liehr, unpublished data].

> **To characterize CG-CNVs of chromosome 19 centromeric region at least probe D19Z3 should be applied. Possibly centromere-near/ centromere-flanking probes could be used as well [Liehr et al., 2006] to exclude small pericentric inversions.**

5.1.2.20. Chromosome 20

- Regular size of centromeric α-satellite DNA D20Z2 (formerly p3.4, pZ20 or X58269): 1,020 kb [Rocchi et al., 1990; Choo et al., 1991; Lo et al., 1999; Alexandrov et al., 2001]
- Cytoband: 20p11.1
- Regular size of centromeric α-satellite DNA D20Z1: 3,900 kb [Bassi et al., 2000]
- Cytoband: 20q11.1
- Centromere size according to GRCh37/hg19: 3,800 kb (see Table 4)

20cen+ and 20cen− CG-CNVs have only been detected so far using probe D20Z2 and not D20Z1, which hybridizes to centromeric region of chromosomes 2 and 20 [Liehr, unpublished data] (Color plate 4, Figure VIII). Thus, only amplification of 20p11.1 and not 20q11.1 was observed. However, heterochromatic enlargement of 20p11.1 [Fryns et al., 1988; Park and Rawnsley, 1996] or 20q11.1 [Petersen, 1986, Romain et al., 1991] has been described cytogenetically (Figure 8B). Furthermore cytoband 20q11.21 has been proven to harbor at least 30 kb of satellite II DNA (GAATG) [Warburton et al., 2008].

> To characterize CG-CNVs of chromosome 20 centromeric region at least the probe D20Z2 should be applied. Possibly centromere-near/centromere-flanking probes could be used as well [Liehr et al., 2006] to exclude small pericentric inversions.

5.1.2.21. Chromosome 21

- Regular size of centromeric α-satellite DNA D21Z1 (formerly αRI or D29750): 420 to 3,150 kb [Lo et al., 1999; Trowell et al., 1993; Marçais et al., 1993; Alexandrov et al., 2001]
- Cytobands: 21p11.1 to 21q11.1
- Centromere size according to GRCh37/hg19: 3,400 kb (see Table 4)

The centromeric sequence D21Z1 is shared by chromosomes 13 and 21. As reported for chromosome 13 this sequence tends to be polymporphic on chromosome 21 (see section 5.1.2.13), but also on other chromosomes (see section 5.1.2.25). There were reports on how to distinguish between chromosomes 13 and 21 (D13Z1 and D21Z1) by primed in situ hybridization (PRINS), however the selected sequences were also heteromorphic and differentiated only 80% of chromosomes 21 clearly [Yang et al., 2001]. FISH can also be applied for sequences distinguishing at a similar efficiency [Soloviev et al., 1998].

More detailed information about the pericentromere of chromosome 21 and four other small alphoid stretches D21Z3, D21Z4, D21Z5, and D21Z7 in the short arm adjacent to D21Z1 can be found in literature [Choo et al., 1991; Alexandrov et al., 2001; Laurent et al., 2003]. Satellite I DNA (pTRI-6) and satellite III DNAs (D21Z6 and another denominated as pTRS-2) also were located there [Lee et al., 1997]. In all acrocentric centromeres there are sequences that are homologous to some DNA stretches located in 5qter, as an early subtelomeric probe showed (cosmid B22a1) [Knight and Flint, 2000].

The prevalence of 21cen– CG-CNVs is 3.7% [Lo et al., 1999; Bossuyt et al., 1995]; the latter refers to complete absence of the centromeric signal. Verma, Batish et al. (1997) point out that approximately an additional 20% of chromosomes 21 have reduced signal intensities. About 13% show enlarged signals after FISH [Verma, Batish et al., 1997]. 21cen– CG-CNVs were also reported by others [Iurov et al., 1991; Mizunoe and Young, 1992; Nilsson et al., 1997; Verma, Batish et al., 1997; Marzais et al., 1999, Liehr et al., 1999; Tepperberg et al., 2001], as were 21cen+ CG-CNVs [Iurov et al., 1991; Verma, Batish et al., 1997; Nilsson et al., 1997].

Structural or size variants on a chromosome 21 centromeric region may also be due to insertion of D15Z1 material (see section 5.1.1.1.2.1).

To characterize CG-CNVs of chromosome 21 centromeric region at least probe D21Z1 should be applied, which also gives signals on chromosome 13. Possibly centromere-near/centromere-flanking probes could be used as well [Liehr et al., 2006] to exclude small pericentric inversions; a short arm probe for all acrocentric chromosomes is helpful in many cases too (section 5.1.1).

5.1.2.22. Chromosome 22
- Regular size of centromeric α-satellite D22Z4: n.d. [Shiels et al., 1997]
- Cytobands: 22p11.2 to 22p11.1
- Regular size of centromeric α-satellite D22Z1 (formerly αXT or D14Z9 or M22273): 2,000 to 2,600 kb [Trowell et al., 1993; Shiels et al., 1997; Lo et al., 1999; Alexandrov et al., 2001]
- Cytobands: 22p11.1 to 22q11.1
- Regular size of centromeric α-satellite D22Z21: 120 to 2,800 kb [Warburton et al., 1986; McDermid et al., 1986]
- Cytoband: 22q11.1
- Centromere size according to GRCh37/hg19: 5,700 kb (see Table 4)

To characterize region 22p11.1–q11.1 the alphoid stretch D22Z1 can be used, which at the same time stains the centromere of chromosome 14; still 22cen+ and 22cen– CG-CNVs [Liehr et al., 1992 and 1998] can be defined using this probe. The alphoid region D22Z4 may vary in size too (Color plate 1, Figure II). According to Antonacci et al. (1995) overall there are four different alphoid stretches on chromosome 22 called D22Z1, D22Z3 [Müllenbach et al., 1996], D22Z4, and D22Z2 [Lee et al., 1997; Choo et al., 1991], and there may be some more [Eisenbarth et al.,1999]. Additionally, band 22q11.1 harbors at least 10 kb of satellite DNA repeat CAGCn and a hsatII repeat of at least 15 kb in size [Warburton et al., 2008]. In all acrocentric centromeres there are sequences that are homologous to some DNA stretches located in 5qter, as a first-generation subtelomeric probe showed (cosmid B22a4) [Knight and Flint, 2000].

Finally, variants on a chromosome 22 centromeric region may be due to insertion of D15Z1 material (see section 5.1.1.1.2.1).

To characterize CG-CNVs of chromosome 22 centromeric region at least probe D14/22Z1, which also gives signals on chromosome 14, and additionally D22Z4 should be applied. Possibly centromere-near/centromere-flanking probes could be used as well [Liehr et al., 2006] to exclude small pericentric inversions; also a short arm probe for all acrocentric chromosomes is helpful in many cases (section 5.1.1).

5.1.2.23. X-Chromosome

- Regular size of centromeric α-satellite DNA DXZ1 (formerly pBAMX7, pXBR-1; pYAM10-40 or X02418): 1,380 to 3,730 kb [Yurov et al., 1987; Choo et al., 1991; Lo et al., 1999; Alexandrov et al., 2001]
- Cytobands: Xp11.1 to Xq11.1
- Centromere size according to GRCh37/hg19: 2,500 kb (see Table 4)

Besides the alphoid DNA stretch DXZ1 the chromosomal band Xp11.1 also harbors hsat4 repeats of at least 67 kb [Warburton et al., 2008]. A γ-satellite DNA (1,205 bp repeat units) covers approximately 50 to 500 kb in pericentromere of the X-chromosome [Lee et al., 1997; Lee et al., 2000; Schueler et al., 2001]. The region between DXZ1 and the euchromatic region in the short arm of the X-chromosome also was studied in detail [Schueler et al., 2001]. Certainly DXZ1 is also present in Xq11.1.

The centromeric region of the X-chromosome is prone to heteromorphisms as the corresponding regions of the 22 autosomes (see sections 5.1.2.1 to 5.1.2.22 and 5.1.2.25). Xcen+ [Liehr et al. 2001 or 2002] and

Xcen– CG-CNVs [Tsuchiya et al., 2001; Tepperberg et al., 2001; Griffiths et al., 1996] are reported using probe DXZ1 [Thilaganathan et al., 2000]. Xcen– CG-CNVs can be so small that a loss of an X-chromosome can be suspected [Tsuchiya et al., 2001; Griffiths et al., 1996].

> **To characterize CG-CNVs of the X-chromosome centromeric region at least probe DXZ1 should be applied. Possibly centromere-near/centromere-flanking probes could be used as well [Liehr et al., 2006] to exclude small pericentric inversions.**

5.1.2.24. Y-Chromosome

- Regular size of centromeric α-satellite DNA DYZ3 (formerly called cos Y84, 3E7 or 853): 285 to 1,020 kb [Lee et al., 1997; Lo et al., 1999] or 5,500kb [Tyler-Smith and Brown, 1987; Waye et al., 1987]
- Cytobands: Yp11.1 to Yq11.1
- Centromere size according to GRCh37/hg19: 1,800 kb (see Table 4)

The DNA stretch DYZ3 has, in contrast to all other centromeric α-satellites, only 170 bp repeats. Besides DYZ3, another Y-chromosome-specific α-satellite DNA called YII3.1 was reported in 1993 [Lee et al., 1997] and bands Yp11.1 and Yq11.1 were proven to harbor at least 100 and 71kb of satellite II DNA (GAATG), respectively [Warburton et al., 2008]. Satellite III DNA of 400 kb length was described at pericentromere of the Y-chromosome [Lee et al., 1997], as well as ß-satellite in Yp11.1 to Yq11.1 [Meneveri et al., 1993; Cooper et al., 1993]. Also ß-satellite and other satellite DNAs were detected in Yq12 [Warburton et al., 2008] (see section 5.1.3.1).

Ycen+ and Ycen– CG-CNVs [Tepperberg et al., 2001] using probe DYZ3 were seen only rarely. However, for the Y-chromosome a frequent inversion variant exists: inv(Y)(p11.2q11.2) [Brothman et al., 2006]. The appearance of such a variant in South Africa was traced back in one study to an Indian origin [Spurdle and Jenkins, 1992].

> **To characterize CG-CNVs of the Y-chromosome centromeric region at least probe DYZ3 should be applied. Possibly centromere-near/centromere-flanking probes could be used as well [Liehr et al., 2006] to exclude small pericentric inversions.**

5.1.2.25. Insertion of Pericentric Material into Other Chromosomes

In sections 5.1.2.1 to 5.1.2.24 all human centromeres and their CG-CNVs as present on their native chromosome are listed. In this section more rare, and maybe underreported variants are summarized; those are due to insertion or translocation of (peri-)centromeric material to other chromosomes and/or their centromeres. Insertion of such (peri)centric material into the (peri)centromeres of other chromosomes should be possible for each α-satellite region, but currently only the following have been reported:

- 1q12 has been inserted in Xq21 in a mother and her daughter [Vust et al., 1998].
- The α-satellite region D4Z1 can be transferred on a chromosome 15 [Liehr, unpublished data].
- Region 9q12 may be inserted in the centromeric region of a chromosome 5 [Fineman et al., 1989; Doneda et al., 1998].
- D13/21Z1 stretches were included in centromeres of chromosome 14 [Lapidot-Lifson et al., 1996], 15 [Acar et al., 2002], or 22 [Verlinsky et al., 1995; Blancato, 1996; Tardy and Tóth, 1997; Bartsch et al., 1993; Tepperberg et al., 2001; Acar et al., 2002]. Here translocations of the entire short arm may be causative.
- D18Z1 has been seen to be inserted in pericentromere regions of chromosome 1 [Musilova et al., 2008], 2 [Collin et al., 2009], 9 [Wei et al., 2007], or 22 [Thangavelu et al., 1998].
- DXZ1 may be present on a chromosome 19 [Winsor et al., 1999], or even on more chromosomal regions simultaneously like 1, 12, and 17 [Liehr et al., 2002].

  Note: In interphase diagnostics, a cross hybridization cannot be distinguished from a complete or partial trisomy; the presence of a heterochromatic sSMC may also be indicated [Liehr et al., 2001].

The origin of additional centromeric signals should be checked carefully. Whenever possible, interphase data should be amended by metaphase FISH.

5.1.2.26. MG-CNVs of the pericentric regions

MG-CNVs in the satellite DNA stretches are underdiagnosed, as normally they are not targeted with any of the current applied cytogenetic, molecular cytogenetic, or molecular tools. Apart from studies on centromeric and pericentric satellite DNAs in the 1970s to the 1990s [Waye and Willard, 1989; Choo et al., 1991; Haaf and Willard, 1992; O'Keefe et al., 1997 and 2000] no large scale or systematic studies of this "undiscovered land" in the human genome have been done. As some satellite DNAs are transcribed into RNA under certain circumstances [Rizzi et al., 2004; Madon, 2005; Enukashvily et al., 2007; Eymery et al., 2010] future studies may be indicated for these regions.

5.1.3. Gonosomal-Derived Heterochromatin

5.1.3.1. Size Variations of Yq12

- Regular size of classical satellite III DNA DYZ1 (formerly pJA1143): 6,800 to 13,600 kb [Rahman et al., 2004]
- Cytoband: Yq12
- Regular size of classical satellite III DNA DYZ2 (formerly pHY 2-1): 4,200 kb [Schmid et al.,1990; Manz et al., 1992]
- Cytoband: Yq12
- Size according to GRCh37/hg19: 30,573 kb (see Table 5)

Two satellite III DNA stretches, DYZ1 and DYZ2, were detected in Yq12; they should be arranged in blocks and in a relation of 2:1 for DYZ1 and DYZ2 [Manz et al., 1992]. Band Yq12 harbors at least 100 kb of satellite II DNA (GAATG) [Warburton et al., 2008] and overall 300 to 10,000 kb of repetitive DNA [Lapidot-Lifson et al., 1996].

Band Yq12 may vary from almost absent to a size that leads to a Y-chromosome of the size of a normal chromosome 13 [Gardner and Sutherland, 2004]; several attempts were done to correlate the length of this heteromorphic region to any phenotypic sign of the corresponding male, which basically all failed [Wall et al., 1988]. Yqh– CG-CNVs have a frequency of 0.09% in the male population, and Yqh+ ones are even rarer according to Gardner and Sutherland (2004). However, as listed in Table 1, Yqh– can be found in 0.78% of the cases. This reflects two things: variations in different populations (0.14% Yqh+ in Scotland [Buckton et al., 1980] versus 83.85% in Eastern India [Verma et al., 1983]) and the nonuniformity of the classification of this kind of CG-CNVs.

Hsu et al. (1987) suggested to define a Yq+ as larger than a chromosome 18 and Y– as smaller than a G-group chromosome of the same metaphase; Verma et al. (1983) introduced the quotient of a Y-chromosome divided by the average length of chromosomes 19 and 20 as an orientation-huge abbreviations from a quotient 1 were considered as Yq– or Yq+ then. However, the suggestion made in Figure 4 to use 16p and 18p as references for all four heterochromatic blocks of human (1q12, 9q12, 16q11.2, and Yq12) seems more feasible. As for enlargements of chromosomes 1 and 9, also extrabands may appear using banding cytogenetics in Yq12 [Lin et al., 1988] (see sections 5.1.2.1, 5.1.2.9, and 5.1.3.16). Up to three [Wall et al., 1988] or even four extrabands [Knuutila and Grippenberg, 1972] were reported. Yqh+ and Yqh– CG-CNVs can be visualized and distinguished from other variants (see section 5.1.1.1.3) by the Yq12 specific classical satellite III probes DYZ1 [Fernández et al., 2001] or DYZ2. While most of Yqh+ and

Yqh− CG-CNVs are inherited, there are three cases with mitotically derived size variations in Yq12 [Akkari et al., 2005].

Finally, an aberrant Yq12 band can be a hint on a der(Y) t(Y;acro)(q12;p12), which is another kind of CG-CNV (section 5.1.1.1.3). However, other gene-rich chromosomal material may be added to Yq12 (e.g., [Gunel et al., 2008; Vialard et al., 2009]; Color plate 2, Figure Va-2).

To distinguish a CG-CNV of Yq12 from a possibly meaningful translocation event at least probe DYZ1 together with a probe for the acrocentric short arms should be applied.

5.1.3.2. Addition of Yq12 to Other Chromosomal Ends
Harmless addition of Yq12 material can also happen to other chromosomes, similar to its addition to acrocentric short arm material (section 5.1.1.1.3). If this event does not go together with a deletion of genetic relevant material on the new formed derivative chromosome this is a CG-CNV without any clinical impact for the carrier. Addition of the band Yq12 to the short arm of an acrocentric chromosome is most often observed (see section 5.1.1.1.2.2) [Gardner and Sutherland, 2004]. There are reports on (potentially) harmless translocations to Xp22.33 [Fernández et al., 1994] or 1p36 [Serakinci et al., 2001; Bucksch et al., 2012]; however, they may be detected due to infertility.

5.1.3.3. Insertion of Yq12
Insertion of Yq12-chromosome material in another chromosome is a rarely reported event: there is one publication on a familial insertion of Yq12 in 1q12 [Sala et al., 2007] and another one on a Yq12 insertion into 11q24 transmitted at least over four generations [Ashton-Prolla et al., 1997]. Also Rahman et al. (2004) report insertions into centromeric regions of chromosome 11 or 15. Due to the lack of corresponding techniques, other comparable cases might not have been reported yet [Spak et al., 1989].

5.1.3.4. MG-CNVs of Yq12
Similar facts as stated earlier for MG-CNVs of the pericentric regions are valid for MG-CNVs in Yq12 (see section 5.1.2.26).

5.1.4. Heterochromatic sSMC

Over 530 autosomal derived sSMC cases are available, which are characterized in detail for their size and genetic content [Liehr, 2013] (Table 8). All these cases are not associated with any clinical abnormalities.

In 344 (i.e., 65%) of these sSMC cases no euchromatic material was present. However, the rates of cases with and without euchromatin vary

Table 8 Well-Characterized sSMC without Clinical Impact on the Carriers (summarized from [Liehr, 2013])*

Chromosome	Cases with euchromatin	Cases without euchromatin
1	5	13
2	10	2
3	10	3
4	1	0
5	10	8
6	1	2
7	2	0
8	10	2
9	16	0
10	7	1
11	2	2
12	6	2
13	1	0
14	6	71
15	40	152
16	11	10
17	2	1
18	9	2
19	5	1
20	6	6
21	10	1
22	17	65
X	0	2
Y	2	0
Overall	**189**	**344**

*Sorted by chromosome and broken down to cases with and without euchromatin.

from chromosome to chromosome (Table 8). In general, in acrocentric-derived sSMCs the cases without euchromatin come in superior numbers (exception chromosome 21), whereas it is the other way round in most nonacrocentric derived sSMCs (exception chromosome 1). Many of the heterochromatic sSMCs are segregating within families, thus, parental studies are indicated here.

The origin of very small sSMCs is best to be established by centromeric probes; using α-satellite probes for chromosomes 15, 14/22, 1, 16, 20, and 5 (see frequencies from Table 8) 94% of such pure heterochromatic sSMCs can be unambiguously characterized.

5.2. EUCHROMATIC CG-CNVS

Here euchromatic CG-CNVs are treated including UBCAs (Chapter 2, section 2.4), sSMC (Chapter 2, section 2.5), EVs (Chapter 2, section 2.6), and gonosomal-derived euchromatin (Chapter 2, section 2.7). Common to all of those euchromatic CG-CNVs is that large, (molecular) cytogenetically visible, euchromatic parts of the genome may be present in copy numbers other than two, without having adverse impact on the health of the human carrier. In other words, most likely the corresponding regions do not contain dosage-dependant genes.

Note: "No clinical signs" can include in this section the following conditions: the carrier has no clinical problem at all, he or she is cytogenetically studied due to infertility, or, especially for UBCAs, the carrier might have consistently mild consequences, much less severe than to be expected for the sheer size of an observed imbalance [Barber, 2005].

It goes without saying that heterochromatic regions of the human genome (see section 5.1) are basically not of interest in this chapter. However, euchromatic insertion in 9q12 was reported [Hou and Wang, 1995; Macera et al., 1995] and should be possible for 1q12, 16q11.2, or Yq12 as well. Also euchromatic translocation to an acrocentric short arm (Table 7) and/or Yq12 is possible (see section 5.1.3.2 and Color plate 2, Figure Va-2). As some of the aforementioned rearrangements were inert for the carrier a possible influence of heterochromatization is discussed, but still remains unclear [Kleinjan and van Heyningen, 1998, Barber, Zhang et al., 2006].

More frequently euchromatic CG-CNVs form as structural and/or numerical chromosomal rearrangements including unbalanced translocations, insertions, direct and indirect duplications, deletions, ring chromosomes, sSMC, and complex rearrangements. All of those aberrations lead as final consequence for the genome to duplications (see section 5.2.1) or deletions (see section 5.2.2).

Thus, everything reported by this point as UBCAs, EVs, and euchromatic sSMCs without clinical consequences (including autosomes and gonosomes) is summarized in the following chromosome-specific sections on euchromatic CG-CNVs. For the first three groups original publications are referred only in exceptional cases, as there are excellent, freely available reviews on these topics [Barber, 2005; Barber, Brasch-Anderson et al. 2013; Liehr, 2012 and 2013]. Also included are subtelomeric polymorphisms, which can be mainly detected by molecular (cyto)genetics [Brown et al., 1990; Wilkie et al., 1991], as are MG-CNVs that can enlarge to sizes visible in cytogenetics. In addition, CNVs detectable by aCGH as well as by FISH are mentioned in this chapter where appropriate.

5.2.1. Euchromatic Gain without Clinical Consequences

Euchromatic CG-CNVs without clinical consequences may be due to intrachromosomal duplication, insertion, presence of an sSMC, or another unbalanced chromosomal rearrangement leading overall to partial gain of copy numbers.

5.2.1.1. Chromosome 1

According to Barber (2005) and Barber, Brasch-Anderson et al. (2013) four families are reported with euchromatic CG-CNVs without any clinical consequences along different regions of chromosome 1. Less severe clinical problems than expected (i.e., slight mental retardation) were observed in three additional pedigrees. All three regions in the short and also the four regions in the long arm of chromosome 1 did not show any overlaps.

According to observations from sSMCs [Liehr, 2013] pericentric regions that may be duplicated without any associated clinical problems span 1p11.2 to 1q12~21.1. The latter goes well together with other data from the literature [Rosenfeld et al., 2012] (for summary see Figure 17 and Table 9). Interestingly, some of the large tandemly repeated DNA families summarized by Warburton et al. (2008) are located with in 1p21.1, 1q21.1, and 1q21.3 and colocalize with regions inert for duplications (see also the appendix).

> **Gain of copy numbers: To characterize euchromatic CG-CNVs of chromosome 1 at present only centromere-near/centromere-flanking probes can be recommended [Liehr et al., 2006].**

5.2.1.2. Chromosome 2

For chromosome 2, large gains of copy number with no or minor clinical signs are scarce. According to sSMC-research the pericentric dosage independent region in the short arm covers 2p11.2 (molecular confirmed: ~89.6 Mb) to the centromere and continues into the long arm up to cytoband 2q11.2 (molecular confirmed: ~101.6 Mb) [Liehr, 2013].

Moreover, a partial trisomy of 2q11.2 to 2q21.1 (~28.1 Mb) due to an insertion into chromosome 8 lead only to minor clinical signs in the carriers [Barber, 2005; Barber, Brasch-Anderson et al., 2013]. There is a recent report on a 3.8 Mb duplication in 2q31.1-q31.2 (position hg18: 175.0-178.6 Mb),

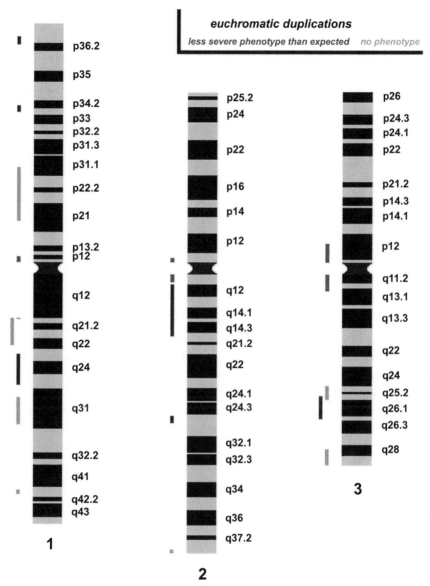

Figure 17 Stretches of euchromatic duplications with no or less severe clinical consequences than expected are symbolized on the left side of each idiogram of chromosomes 1, 2, and 3. The data obtained from sSMC research is highlighted by three vertical stripes in the vertical lines symbolizing gain of copy numbers without associated clinical signs.

Table 9 Summary of Reported Euchromatic CG-CNVs Leading to Gain of Copy Numbers for Chromosome 1

Cytoband	Molecular position (hg 18) [Mb]	Clinical signs	Reference besides [Barber, 2005; Barber, Brasch-Anderson et al., 2013]
chromosome 1			
1p36.33–p36.22	6.3–10.8	mild to moderate	Tonk et al, 2005
1p34.2–p34.1	40.7–44.2	mild to moderate	Bisgaard et al., 2007
1p31–p21	n.a.	none	—
1p11.2–q12∼21.1	115.8–142.4	none	Liehr, 2013
1q11–q22	n.a.	none	—
1q31.1–q32.3	184.5–198.2	none	Mrasek et al., 2008
1q42.11–q42.12	n.a.	none	—
1q23–q25	n.a.	mild to moderate	—

visible at least in molecular cytogenetics, leading to mild to moderate clinical signs in a family with syndactyly and nystagmus [Ghoumid et al., 2011]. Also, the subtelomeric region of chromosome 2 long arm is subject to heteromorphisms including duplications, as discovered by the application of the P1 clone 210E14 and cosmid 2112b2 [Knight and Flint, 2000] (for summary see Figure 17). Two large tandemly repeated DNA families summarized by Warburton et al. (2008) are located with in 2q11.2, and thus colocalize with a centromere-near region inert for duplications (see also the appendix).

> **Gain of copy numbers: For euchromatic CG-CNVs characterization of chromosome 2 at present only centromere-near/centromere-flanking probes can be recommended [Liehr et al., 2006].**

5.2.1.3. Chromosome 3

No clinical signs were associated with a duplication of 3q25 to 3q25 (∼10.4 Mb) or 3q28 to 3q29 (∼8.6 Mb), while a partial trisomy of 3q25.3 to 3q26.2 (∼17 Mb) lead to microcephaly—congenital heart defect and deafness in the corresponding affected families [Barber, 2005; Barber, Brasch-Anderson et al., 2013]. The pericentric uncritical region for gain of copy numbers spans 3p12.2 (molecular confirmed: ∼74.7 Mb) to 3q13.11 (molecular confirmed: ∼104.8 Mb) [Liehr, 2013] (for summary see Figure 17).

Gain of copy numbers: As mentioned for chromosomes 1 and also in chromosome 3 only the euchromatic pericentric CG-CNVs may be characterized applying centromere-near/centromere-flanking probes [Liehr et al., 2006].

5.2.1.4. Chromosome 4

Partial trisomy of band 4p16.1 [Rodriguez et al., 2007] did not lead to any clinical signs. Interestingly, the cytoband 4p16.1 [Balikova et al., 2008] is also reported as an EV region [Barber, Brasch-Anderson et al., 2013]. Furthermore, partial trisomies of 4q31.3 to 4q33 ($\sim$10.7 Mb) [Barber, 2005; Barber, Brasch-Anderson et al., 2013] and 4q32 to 4q35 ($\sim$170 Mb) [Kim et al., 2011] were without clinical problems. The pericentric region inert to gain of copy numbers spans 4p13 (molecular confirmed: $\sim$44.0 Mb) to 4q13.1 (molecular confirmed: $\sim$58.7 Mb) [Liehr, 2013].

Gain of copy numbers in the region 4p16.3, which if deleted leads to Wolf-Hirschhorn syndrome, was repeatedly reported to be associated with only slight clinical features [Carmany and Bawle, 2011] as was gain of 4p13 to 4p11 [Liehr et al., 2011]. Moreover, partial trisomies of 4q31 to 4qter in general were associated with only mild phenotypic anomalies, like slight (mental) retardation and/or dysmorphism [Barber, 2005; Barber, Brasch-Anderson et al., 2013] (for summary see Figure 18 and Table 10). Even a family with triplication of 4q32.1-q32.2 (position hg18: 157.6–161.6 Mb) is reported with comparatively minor clinical features [Wang et al., 2009]. This observation fits well with numerous cases with Cri du Chat syndrome and in parts large duplications on 4qter, which did not affect the severity of the corresponding 5p-syndrome [Shet et al., 2012].

Gain of copy numbers: Chromosome 4 has several euchromatic CG-CNVs—centromere-near/centromere-flanking probes [Liehr et al., 2006], probes for the EV in 4p16.1 (e.g. RP11-637J21, RP11-751L19, RP11-264E23 and/or RP11-180A12 [Balikova et al., 2008]), and such for the terminal region of chromosome 4q can be helpful here.

5.2.1.5. Chromosome 5

A duplication 5q15 to 5q21 ($\sim$16.3 Mb) was reported to be an euchromatic CG-CNV in one family. Only slight (mental) retardation was seen in the case of a pure dup(5)(q15q22.1) ($\sim$13.6 Mb), and when dup(5)(pterp15.1) ($\sim$15 Mb) or a dup(5)(q35qter) ($\sim$3.4 Mb) was combined with a del(4)(q34) or a del(10)(q26.13), respectively [Barber,

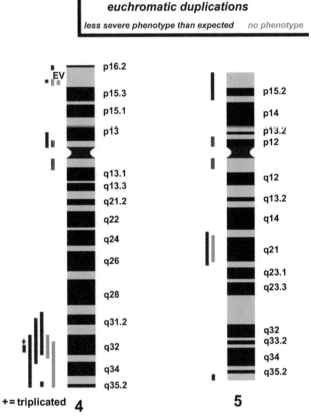

Figure 18 Euchromatic duplications with no or less severe clinical consequences than expected are shown for chromosomes 4 and 5 as described in Figure 17. A euchromatic variant region is highlighted by the abbreviation EV. Also one case with partial triplication is shown (+).

2005; Barber, Brasch-Anderson et al., 2013]. Concerning duplications of the pericentromere of chromosome 5 the uncritical region spans 5p13.1 (molecular confirmed: ~37.2 Mb) to 5q11.2 (~55.3 Mb) [Liehr, 2013] (for summary see Figure 18). There is one report of a duplication 5pter to 5p14 with minor clinical signs; however, there was no molecular confirmation of the additional material to be derived from chromosome 5 [Chia et al., 1987].

Gain of copy numbers: For euchromatic CG-CNVs characterization of chromosome 5 at present only centromere-near/centromere-flanking probes can be recommended [Liehr et al., 2006].

Table 10 Summary of Reported Euchromatic CG-CNVs Leading to Gain of Copy Numbers for Chromosome 4

Cytoband	Molecular position (hg 18) [Mb]	Clinical signs	Reference besides [Barber, 2005; Barber, Brasch-Anderson et al., 2013]
chromosome 4			
4p16.3	n.a.	mild to moderate	Carmany and Bawle, 2011
4p16.1-p16.1	8.5-11.7	none	Rodriguez et al., 2007
4p16.1-p16.1	8.7-9.4	none	Balikova et al., 2008
4p13-p11	39.8 to 48.9	mild to moderate	Liehr et al., 2011
4p13-q13.1	44.0-58.7	none	Liehr, 2013
4q31-qter	n.a.	mild to moderate	Moreira and Riegel, 2000
4q31.1 to 4q32.3	n.a.	mild to moderate	—
4q31.22-q33	n.a.	mild to moderate	—
4q31.3-q33	n.a.	none	—
4q32-q35	n.a.	none	—
4q35.1-q35.2	185.9-188.6	mild to moderate	Bisgaard et al., 2007

5.2.1.6. Chromosome 6

No clinical signs were present if the regions 6q16.1 to 6q21 (position hg18: 95.5-106.7 Mb [Spreiz et al., 2010]), 6q23.3 to 6q24.2 ($\sim$8.1 Mb), or 6q24.2 to 6q24.2 ($\sim$2 Mb) were duplicated. Only slight (mental) retardation was observable in the case of 6q21 to 6q22.1 (5–10 Mb) duplication alone, or a dup(6)(pterp23) ($\sim$17.4 Mb) together with a del(20)(p13) [Barber, 2005; Barber, Brasch-Anderson et al., 2013]. According to the literature, the pericentric uncritical region for duplications spans 6p12.1 (molecular confirmed: $\sim$57.4 Mb) to 6q12 (molecular confirmed: $\sim$65.2 Mb) [Liehr, 2013] (for summary see Figure 19).

Note: Imprinting may cause problems here in form of paternal uniparental disomy (UPD) 6 leading to transient neonatal diabetes (TND; OMIM #601410) – for summary on UPD see Liehr (2010).

Gain of copy numbers: For euchromatic CG-CNVs characterization of chromosome 6 at present only centromere-near/centromere-flanking probes can be recommended [Liehr et al., 2006].

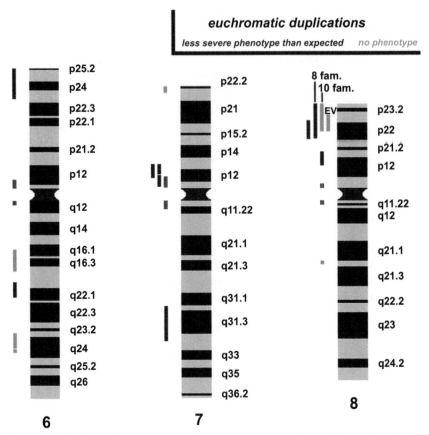

Figure 19 Euchromatic duplications for chromosomes 6 through 8 as described in Figure 17. Abbreviations: EV = euchromatic variant; fam. = families.

5.2.1.7. Chromosome 7

An obviously harmless euchromatic CG-CNV was reported as terminal duplication of 7pter to 7p22.3 (∼3.5 Mb) in a father and his child [Barber, 2005; Barber, Brasch-Anderson et al., 2013]. When the region 7p11.2 (molecular confirmed: ∼55.4 Mb) to 7q11.22 (molecular confirmed: ∼67.0 Mb) is present three times, no clinical abnormalities are expected [Liehr, 2013].

Only minor signs like growth retardation or slight developmental delay were seen in three families with dup(7)(p13p12.2) (∼5.5 Mb), dup(7)(p13p12.1) (∼6.9 Mb), or dup(7)(p12p11.2) (∼6.0 Mb). With respect to the size of the imbalance the duplication of the bands 7q32 to 7q36.1 (∼17.8 Mb) was associated with almost no clinical signs (i.e., developmental delay) or behavioral problems in two generations [Barber, 2005; Barber, Brasch-Anderson et al., 2013] (for summary see Figure 19).

> *Note*: Imprinting may cause problems here as maternal UPD leading to Silver Russel syndrome (SRS; OMIM #180860) [Liehr, 2010].

> **Gain of copy numbers: For euchromatic CG-CNVs characterization of chromosome 7 at present only centromere-near/centromere-flanking probes can be recommended [Liehr et al., 2006].**

5.2.1.8. Chromosome 8

Duplications in the short arm of chromosome 8 with no or only minor clinical consequences are reported in 22 families [Barber, 2005; Barber, Brasch-Anderson et al., 2013]. Absence of clinical signs was reported for partial trisomies 8p23.3 to 8p23.1 ($\sim$6.1 Mb), 8p23.2 to 8p23.2 (position hg 18: 2.5-6.0 Mb [Harada et al., 2002]), 8p23.2 to 8p23.1 ($\sim$4.6 Mb), 8p23.1 to 8p23.1 (3 families $\sim$6.5 Mb), and 8p22 to 8p22 ($\sim$3.4 Mb). However, in 12 of the 22 families with similar stretches of duplications different minor clinical signs were reported, including miscarriages, phenotypic abnormalities, slight (mental) retardation, and congenital heart defects. The affected regions are 8pter to 8p22 ($\sim$17.8 Mb), 8p21.1 to 8p12 ($\sim$6.9 Mb), 8p22 to 8p21.3 or 8p23.1 to 8p22 ($\sim$9.6 Mb), 8p23.1 to 8p22 ($\sim$9.6 Mb), 8p23.1 to 8p21.3 (2 families; 9.6 Mb), 8p23.2 to 8p21.1 (position hg 18: 3.5-10.3 Mb [Glancy et al., 2009]), and 8p23.1 to 8p23.1 ($\sim$3.8 Mb in 2 families, $\sim$6.5 Mb in 3 families) [Barber, 2005; Barber, Brasch-Anderson et al., 2013] (for summary see Figure 19).

> *Note*: The band 8p23.1 was reported to harbor one of the regions described as EVs; 12 families are summarized by Barber (2005) and Barber, Brasch-Anderson et al. (2013) as carriers of such a CG-CNV. This region covers 8p23.1 from position 7.3 to 12.5 Mb (hg18) [Barber et al., 2005] and can be detected using BAC probes like RP11-122N11 and RP11-24D9 (Color plate 4, Figure IXa).

The pericentromere has a stretch inert to duplications that spans 8p11.22 (molecular confirmed: $\sim$42.5 Mb) to 8q11.21 (molecular confirmed: $\sim$48.3 Mb) [Liehr, 2013]. Recently a CG-CNV in 8q21.2 was reported in the region of 86.9 Mb detectable by BAC RP11-96G1 [Manvelyan et al., 2011]. This region harbors segmental duplications, which also can be proven by aCGH as MG-CNV (for summary see Figure 19).

Gain of copy numbers: Chromosome 8 has three identified stretches of euchromatic CG-CNVs—centromere-near/centromere-flanking probes [Liehr et al., 2006] and probes for the EV in 8p23.1 (e.g., RP11-122N11 and RP11-24D9 [Barber, pers. communication]) can be applied.

5.2.1.9. Chromosome 9

A CG-CNV in 9p21.3 to 9p12 (~21 Mb) did not have any clinical consequences in a girl who inherited this duplication due to a maternal insertion [Barber, 2005; Barber, Brasch-Anderson et al., 2013]. The pericentric stretch covering 9p13.1 (molecular confirmed: ~43.0 Mb) to 9q21.12 (molecular confirmed: ~70.5 Mb) is not deleterious in the case of duplication [Liehr, 2013].

Note: Fourteen families are summarized by Barber (2005) and Barber, Brasch-Anderson et al. (2013) presenting enlargement of 9p13.1-p11.2; this potentially duplicated or even deleted region (Figure 16: 9ph–variant; see section 5.2.2.9) is classified as an EV. As shown by Kosyakova et al. (2013) this region is covering at least the positions 38.8 to 42.0 Mb and can be visualized using the BAC probes RP11-402N8 and RP11-128P23, for example (for summary see Figure 20).

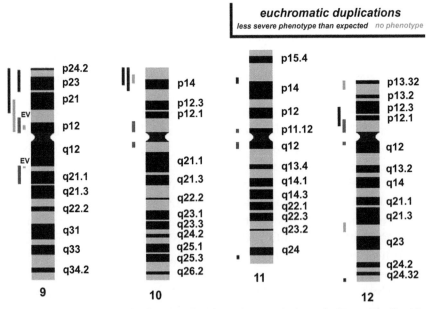

Figure 20 Euchromatic duplications for chromosomes 9 through 12 as described in Figure 17. Abbreviation: EV = euchromatic variant.

Interestingly, Barber (2005) and Barber, Brasch-Anderson et al. (2013) discussed the region 9q13 to 9q21.11 also as a potential EV, which is supported by the studies of Kosyakova et al. (2013) as well; they showed that this region covers at least 70.1 to 71.0 Mb, detectable by BACs RP11-211E19 and RP11-88I18.

For a large dup(9)(p24p22) ($\sim$13.2 Mb) less clinical problems than expected are reported over two generations as developmental and growth delay [Barber, 2005; Barber, Brasch-Anderson et al., 2013]. Also no significant clinical signs were observed in a case with partial trisomy 9pter to 9p13.3 (position hg 18: 0-34.6 Mb) [Bouhjar et al., 2011] (for summary see Figure 20).

> **Gain of copy numbers: The euchromatic CG-CNVs characterization in chromosome 9 is restricted to centromere-near/centromere-flanking regions, which colocalize with EVs; probes as reported in Kosyakova et al. (2013) can be applied: RP11-402N8 and/or RP11-128P23 in 9p and RP11-211E19 and/or RP11-88I18 in 9q.**

5.2.1.10. Chromosome 10

Apart from the duplication of 10p14 to 10p13 ($\sim$5.3 Mb), which was reported to be without clinical impact at all in one family, duplications of 10p11.22 (molecular confirmed: $\sim$34.8 Mb) to 10q11.22 (molecular confirmed: $\sim$43.8 Mb) were without clinical impact as well [Liehr, 2013]. Additionally, dup(10)(p15p13) ($\sim$17 Mb) or duplication 10p15 to 10p14 ($\sim$13 Mb) led to speech delay or multiple congenital abnormalities, however less severe as expected for the size of the aberration [Barber, 2005; Barber, Brasch-Anderson et al., 2013] (for summary see Figure 20).

> **Gain of copy numbers: Euchromatic CG-CNVs in chromosome 10 are restricted to centromere-near/centromere-flanking regions, and possibly parts of its short arm; pericentric probes can be found in genome browsers or elsewhere [Liehr et al., 2006].**

5.2.1.11. Chromosome 11

No clinical signs are associated with partial trisomies of 11p11.12 (molecular confirmed: $\sim$50.5 Mb) to 11q12.2 (molecular confirmed: $\sim$60.2 Mb) [Liehr, 2013]. Apart from that one family with duplication of 11p15.1 to 11p14.3 (3-5 Mb) showed phenotypic abnormalities like developmental retardation [Barber, 2005; Barber, Brasch-Anderson et al., 2013]. Also one patient having a duplication of 11q24.2 to 11q25 (position hg18: 124.4-131.1 Mb) presented only minimal symptoms [Göhring et al., 2008] (for summary see Figure 20).

Note: Imprinting may cause problems here; maternal UPD can be associated with Silver Russel syndrome (SRS; OMIM #180860) and paternal UPD with Beckwith-Wiedemann syndrome (BWS; OMIM #130650) [Liehr, 2010].

Gain of copy numbers: For euchromatic CG-CNVs characterization of chromosome 11 at present only centromere-near/centromere-flanking probes can be recommended [Liehr et al., 2006].

5.2.1.12. Chromosome 12
The region 12p12.1 (molecular confirmed: ~28.4 Mb) to 12q12 (~39.9 Mb) was proven to be insensitive to gain of copy numbers [Liehr, 2013]. Also no phenotypic effect were seen for a dup(12)(q21.32q22) (position hg18: 86.5-92.7) present over two generations [Barber et al., 2007] and for duplication of 12pter to 12p13.31 (position hg18: 0-6.0 Mb) [Madrigal et al., 2012]. However, mental retardation and mild facial features were the result of a dup(12)(p12.3p11.2) (position hg18: 17.6-28.9 Mb) [Liang et al., 2006; Barber, 2005; Barber, Brasch-Anderson et al., 2013], and a duplication of chromosome 12q24.33 (position hg18: 130.1-132.3) combined with a deletion of 17pter (size 1.3 Mb) led to only minor symptoms [Schoumans et al., 2005] (for summary see Figure 20).

Gain of copy numbers: For euchromatic CG-CNVs characterization of chromosome 12 at present only centromere-near/centromere-flanking probes can be recommended [Liehr et al., 2006].

5.2.1.13. Chromosome 13
In chromosome 13, as in all acrocentrics, only the long arm is euchromatic; here the region insensitive to gain of copy numbers spans up to cytoband 13q12.11 (molecular confirmed: ~19.3 Mb) [Liehr, 2013]. Duplications 13q13 to 13q14.3 (~11.6 Mb), 13q14.2 to 13q21.1 (~18.3 Mb), 13q21.1 to 13q21.32 (position hg15: 57.6-68.2 Mb), [Daniel et al., 2007] or the subtelomeric region [Ballif et al., 2000] were not associated with clinical signs [Barber, 2005; Barber, Brasch-Anderson et al., 2013]. Two additional cases with dup(13)(q21.1q21.33) (position hg18: 57-69 Mb) [Lopez-Exposito et al., 2008] and dup(13)(q21.31q31.1) (hg18: 62.2-83.3 Mb) [Mathijssen et al., 2005] showed slight phenotypic signs only [Barber, 2005; Barber, Brasch-Anderson et al., 2013]. A large partial trisomy including 13q22 to 13qter (~280 Mb) was associated with comparatively mild clinical problems [Begovic et al., 1998] (for summary see Figure 21).

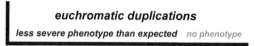

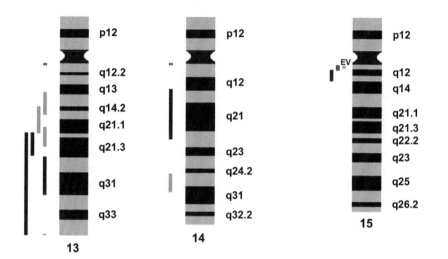

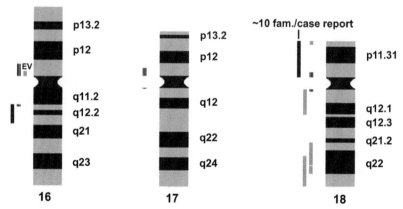

Figure 21 Euchromatic duplications for chromosomes 13 through 18 as described in Figure 17. Abbreviation: EV = euchromatic variant; fam. = families.

Gain of copy numbers: For euchromatic CG-CNVs characterization of chromosome 13 at present only the centromere-near/centromere-flanking probe in the long arm can be recommended [Liehr et al., 2006].

5.2.1.14. Chromosome 14

Euchromatic CG-CNV as gain of copy numbers span the centromeric region of chromosome 14 up to 14q11.2 (molecular confirmed: ~ 20.24 Mb) [Liehr, 2013]. While a dup(14)(q24.3q31) (~ 9.8 Mb) was not correlated with clinical signs, a three-generation family with a dup(14)(q13q22) (~ 26.1 Mb) had developmental delay [Barber, 2005; Barber, Brasch-Anderson et al., 2013] (for summary see Figure 21).

Note: Imprinting may cause problems here, as maternal UPD induces Temple syndrome (TS; see OMIM *605636 and #176270), and paternal UPD induces the "paternal UPD(14) syndrome" (patUPD(14); OMIM #608149) [Liehr, 2010].

Gain of copy numbers: For euchromatic CG-CNVs characterization of chromosome 14 at present only the centromere-near/centromere-flanking probe in the long arm can be recommended [Liehr et al., 2006].

5.2.1.15. Chromosome 15

According to sSMC research the region inert to gain of copy numbers goes up to 15q11.2 ~ 12 (~ 21.18 Mb); here not only duplications but even partial hexasomy of the corresponding region was tolerated [Liehr, 2013]. Five families with intrachromosomal duplication of 15q11 to 15q13 (~ 4 Mb) and one with dup(15)(q11q12) (~ 4 Mb) did not show clinical signs in at least one generation. Notably this subband q11.2 is known as MG-CNV (Figure 2) as well as CG-CNV and additionally is considered an EV (position hg18: 18.5-20.0 Mb); approximately 30 reports on families with an EV in 15q11.2 to 15q13 are available [Barber, 2005; Barber, Brasch-Anderson et al., 2013] (Color plate 4, Figure IXb). Also, a large family with autism spectrum disorder due to duplication in 15q11-q13 (position hg 18: 20.7-26.7 Mb) is reported [Piard et al., 2010] (for summary see Figure 21).

Note: Imprinting may cause problems here as Prader Willi syndrome due to maternal UPD (PWS; OMIM#176270) or Angelman syndrome (paternal uniparental disomy;AS; OMIM #105830) [Liehr, 2010].

Gain of copy numbers: The euchromatic CG-CNVs characterization in chromosome 15 is restricted to the centromere-near/centromere-flanking region in the long arm, which colocalizes with an EV; probes as reported in Liehr et al. (2006) can be applied.

5.2.1.16. Chromosome 16

Chromosome 16 centric region encompasses a large heterochromatic block in 16q11.2, thus, the region not affected by euchromatic gain of copy numbers spans a large stretch, which according to Liehr (2013) is ranging from 16p11.2 (molecular confirmed: ~29.0 Mb) to 16q12.1 (molecular confirmed: ~46.0 Mb). In addition, duplication 16q12.1 to 16q12.1 (~5.1 Mb) is reported to be unproblematic at least in one carrier, and there are two more families with dup(16)(q11.2q12.1) (~5.1 Mb) or a duplication of 16q11.2 to 16q13.1 (~11.9 Mb) present over at least three generations, with just speech or developmental delay [Barber, 2005; Barber, Brasch-Anderson et al., 2013] (for summary see Figure 21).

Note: An EV is mapped to 32.0 to 34.4 Mb (16p11.2-12), which needs to be distinguished from potentially pathological duplications in 16p12--region [Barber, Brasch-Anderson et al., 2013; Barber, Hall et al., 2013]; approximately 15 families are reported [Barber, 2005; Barber, Brasch-Anderson et al., 2013]. To characterize the CG-CNV in 16p11.2 as EV BACs, RP11-96K14 (position hg18: 32.4-32.6 Mb) or RP11-488I20amp (position hg18: 34.3-34.5 Mb) may be applied (Color plate 4, Figure IXc).

Gain of copy numbers: The euchromatic CG-CNVs characterization in chromosome 16 is restricted to centromere-near/ centromere-flanking regions; this region of the p-arm colocalizes with an EV. Centromere-near/centromere-flanking probes [Liehr et al., 2006] and probes for the EV in 16p11.2 (e.g., RP11-96K14 and RP11-488I20 [Barber, pers. communication]) can be applied.

5.2.1.17. Chromosome 17

For chromosome 17 exclusively pericentric duplications were reported to behave as euchromatic CG-CNVs if not exceeding 17p11.2 (molecular confirmed: ~18.7 Mb) to 17q11.2 (molecular confirmed: ~23.3 Mb) [Liehr, 2013] (for summary see Figure 21).

Gain of copy numbers: For euchromatic CG-CNVs characterization of chromosome 17 at present only centromere-near/centromere-flanking probes can be recommended [Liehr et al., 2006].

5.2.1.18. Chromosome 18

Interestingly, partial trisomy of the entire short arm of chromosome 18 leads only to minor signs and symptoms [Moog et al., 2000; Maranda et al., 2006, Dufke et al., 2006; Liehr, 2013], while partial tetrasomy 18p is connected with the so-called i(18p)-syndrome and severe clinical features [Liehr, 2013]. Duplications 18pter to 18p13.3 (~1.9 Mb in size) [Srebniak et al., 2011], 18q11.2 to 18q12.2 (~10 Mb), and 18q21.31 to 18q22.2 (position hg18: 53-65 Mb) [Ceccarini et al., 2007] were reported to be without clinical consequences for their carriers [Barber, 2005; Barber, Brasch-Anderson et al., 2013]. And according to Liehr (2013) the region 18p11.21 (molecular confirmed: ~12.8 Mb) to 18q11.2 (molecular confirmed: ~18.1 Mb) is clinically inert to gain of copy numbers. Recently, also the following cases were reported with partial duplications and no clinical signs: dup(18)(q21.21q22) spanning 65.3 to 69.1 Mb [Henson et al., 2012] and dup(18)(q21.33q23) spanning 58.7 to 76.1 Mb [Quiroga et al., 2011] (for summary see Figure 21).

> **Gain of copy numbers:** For euchromatic CG-CNVs characterization of chromosome 18 at present only centromere-near/centromere-flanking probes can be reliably recommended [Liehr et al., 2006].

5.2.1.19. Chromosome 19

Only pericentric stretches are yet reported as euchromatic CG-CNVs. The corresponding regions were narrowed down to 19p13.11 (molecular confirmed: ~17.5 Mb) to 19q13.11 (molecular confirmed: ~43.88 Mb) [Liehr, 2013] (for summary see Figure 22).

> **Gain of copy numbers:** For euchromatic CG-CNVs characterization of chromosome 19 at present only centromere-near/centromere-flanking probes can be recommended [Liehr et al., 2006].

5.2.1.20. Chromosome 20

The region 20p11.22 (molecular confirmed: ~22.8 Mb) to 20q11.21 (molecular confirmed: ~29.9 Mb) was shown to be inert against gain of copy numbers in terms of clinical outcome of the carrier [Liehr, 2013]. Also a case with a dup(20)(p12.2p12.1)mat (position hg18: 11.7–16.7 Mb) was observed with normal karyotype (Color plate 4, Figure Xa; for summary see Figure 22).

> *Note:* Imprinting may cause problems here by maternal or paternal UPD leading to the corresponding maternal UPD(20) syndrome or pseudo-hypoparathyroidism (PHP OMIM #103580, #603233, #612462) [Liehr, 2010].

> **Gain of copy numbers:** For euchromatic CG-CNVs characterization of chromosome 20 at present only centromere-near/centromere-flanking probes can be recommended [Liehr et al., 2006].

5.2.1.21. Chromosome 21

For acrocentric chromosome 21 the heterochromatic short arm can form heterochromatic CG-CNVs (section 5.1.1). Euchromatic CG-CNVs are just reported for 21q11.1 to 21q11.2~21.1 (molecular confirmed: ~16.9 Mb)

[Liehr, 2013]. Besides, duplication 21q22 to 22qter (~ 18.3 Mb) leads to only slight phenotypic abnormalities [Barber, 2005; Barber, Brasch-Anderson et al., 2013] (for summary see Figure 22).

> **Gain of copy numbers: For euchromatic CG-CNVs characterization of chromosome 21 at present only the centromere-near/ centromere-flanking probe in the long arm can be recommended [Liehr et al., 2006].**

5.2.1.22. Chromosome 22

The centromere-near region not leading to clinical problems in chromosome 22q spans 22q11.1 to 22q11.21 (molecular confirmed: ~ 16.37 Mb) [Liehr, 2013]. Dicentric variants of chromosome 22 leading to a euchromatic trisomy 22q11.21 may also be considered as a CG-CNV (Color plate 4, Figure Xb), as observed in three families [Liehr, unpublished data]

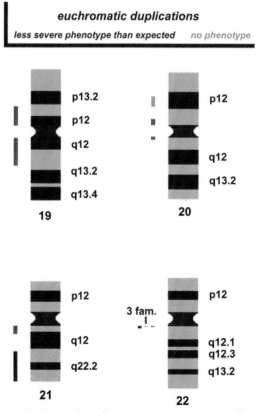

Figure 22 Euchromatic duplications for chromosomes 19 through 22 as described in Figure 17. Abbreviation: fam. = families.

(for summary see Figure 22). If duplication becomes slightly larger cat eye syndrome features may appear [Knijnenburg et al., 2012].

> **Gain of copy numbers: For euchromatic CG-CNVs characterization of chromosome 22 at present only the centromere-near/centromere-flanking probe in the long arm can be recommended [Liehr et al., 2006].**

5.2.1.23. X-Chromosome

The gonosome X is one of the gene-richest human chromosomes. As mentioned in Chapter 2, section 2.7.2.2, here the Lyon hypothesis and almost complete genetic inactivation has to be considered in case of additional copies of the X-chromosome. The latter is valid even for enlarged, duplication-carrying X-chromosome variants, as long as the XIST region is present on them. Conditions like triple-X syndrome [Otter et al., 2012] and Turner syndrome-like phenotypes due to partial gain (and simultaneous partial loss) of X-chromosome material [Liehr et al., 2007] might be carefully considered as CG-CNV too. On the other hand, no euchromatic duplication CG-CNVs were yet reported without considering at the same time X-chromosome inactivation as reason for a normal clinical phenotype of the carriers. Not even small pericentric duplications due to sSMC presence was reported yet [Liehr, 2013].

The only variant region known for the X-chromosome was covered by the cosmid CY29, which covers the subtelomeric regions Xpter and Ypter [Knight and Flint, 2000].

> **Gain of copy numbers: Any probe determining the copy number of the X-chromosome in meta- or interphase can be applied.**

5.2.1.24. Y-Chromosome

The male-specific gonosome is the gene-poorest of human chromosomes. Unlike the X-chromosome, one additional Y-chromosome is tolerated almost without any clinical impact [Stochholm et al., 2012]. Also direct duplication of a Y-chromosome without clinical impact was reported [Kuan et al., 2012]. Even partial tri- to tetrasomy of the Y-chromosome is possible, often combined with Turner syndrome-like karyo- and/or phenotypes [Liehr et al., 2007] (see Chapter 2, section 2.7.2.1). Also the subtelomeric region in Ypter (and Xpter) covered by the cosmid CY29 behaves heteromorphic [Knight and Flint, 2000].

> **Gain of copy numbers: Any probe determining the copy number of the Y-chromosome in meta- or interphase can be applied.**

5.2.2. Euchromatic Loss without Clinical Consequences

Euchromatic CG-CNVs without clinical consequences may be due to interstitial or terminal deletions, including ring chromosome formation, leading to partial loss of copy numbers.

5.2.2.1. Chromosome 1

Slight phenotypic abnormalities like developmental delay were associated in one family, each with a deletion of 1p32.1 to 1p31.3 (position hg18: 59.0–64.6 Mb) [Bisgaard et al., 2007] or 1q42.1 to 1q42.3 (~7.1 Mb) [Barber, 2005; Barber, Brasch-Anderson et al., 2013]. Apart from that, a deletion of 1p34.3 to p34.1 (~9 Mb) [Martínez et al. 1999] and another one from 1q44 to 1qter (0.4 Mb) [Queralt et al., 2008] had either no or only minor clinical impact (for summary see Figure 23).

> **Loss of copy numbers: No probe can be recommended yet for euchromatic CG-CNVs of chromosome 1.**

5.2.2.2. Chromosome 2

Several regions may be deleted in chromosome 2 without having clinical consequences: 2p12 to 2p12 (2 families: ~6.1 Mb and ~6.9 Mb), 2p12 to 2p11.2 (~7.5 Mb), 2q13 to 2q13 (position hg18: 111.2–112.8 Mb [Bisgaard et al., 2007]), 2q13 to 2q14.1 (position hg18: 114.0–117.4 Mb) [Barber, Maloney et al., 2006], and 2q14.1 to 2q14.2 (117.3–119.4 Mb) [Barber, Maloney et al. 2006; Barber, 2005; Barber, Brasch-Anderson et al., 2013]. Finally, the subtelomeric region 2qter covered by P1 clone 210E14 or cosmid 2112b2 was shown to be heteromorph and the FISH probe to be deleted in normal persons [Knight and Flint, 2000] (for summary see Figure 23).

> **Loss of copy numbers: No probe can be recommended yet for euchromatic CG-CNVs of chromosome 2. Maybe one for cytoband 2p12 could be helpful in some cases.**

5.2.2.3. Chromosome 3

In chromosome 3 the region 3pter to 3p25(.3) may be absent without clinical consequences (3 families: ~9.1 to ~10.1 Mb); also the same region was absent in two families with slight phenotypic abnormalities (2 families: ~8.8 Mb and ~9.1 Mb). Moreover, one family with an unaffected and two affected family members and del(3)(p26.3) is known, for which the two-hit-model [Girirajan et al., 2010] might be valid [Cuoco et al., 2011]. One family was reported with a del(3)(p12.3p11.2)

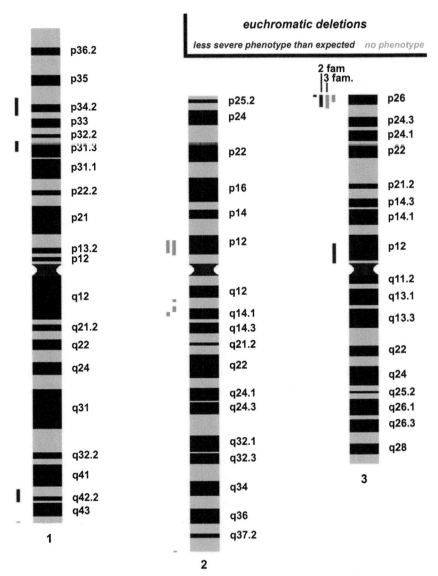

Figure 23 Stretches of euchromatic deletions with no or less severe clinical consequences than expected are symbolized on the left side of each idiogram of chromosomes 1, 2, and 3. Abbreviation: fam. = families.

(~10 Mb) with no/minimal clinical signs like learning difficulties [Barber, 2005; Barber, Brasch-Anderson et al., 2013] and another one with a del(3)(p26.1) and no clinical symptoms (Color plate 4, Figure Xia; for summary see Figure 23).

Loss of copy numbers: A probe for 3p26 to 3pter (e.g., commercially available subtelomeric) can be recommended for euchromatic CG-CNVs of chromosome 3.

5.2.2.4. Chromosome 4

Even though the short arm of chromosome 4 harbors the Wolf-Hirschhorn-microdeletion-syndrome critical region there are several reports on families with deletions in these chromosomal subbands with (almost) healthy family members [Barber, 2005; Barber, Brasch-Anderson et al., 2013]. Furthermore, there is one case with a deletion 4q34.1 to 4q34.3 (~9.4 Mb) and no clinical signs. In parts, surprisingly minor clinical signs are reported for deletions in 4q32 to q33 to 4qter, including one case with del(4)(q34) (~13.3 Mb) and dup(5)(pterp15.1), and another one with del(4)(q35.2) (~1.3 Mb) and del(22)(q11.2) (1 family, each) [Barber, 2005; Barber, Brasch-Anderson et al., 2013]. Also two of my unpublished observations are included in Table 11 and Figure 24: a clinically healthy mother had a child with developmental delay and both had a del(4)(q34.1~34.2) (Color plate 4, Figure XIb), while no clinical signs were observed in a person with del(4)(q35.1) [Liehr, unpublished data]. The findings summarized in Figure 24 are also supported by only minor clinical findings detected in a female with ring chromosome 4 and terminal deletions in 4pter (815 kb) and 4qter (150 kb) [Lee et al., 2005]. To the best of our knowledge no deletion in the EV in 4p16.1 has been reported.

Table 11 Summary of Reported Euchromatic CG-CNVs Leading to Loss of Copy Numbers for Chromosome 4

Cytoband	Molecular position (hg 18) [Mb]	Clinical signs	Reference besides [Barber, 2005; Barber, Brasch-Anderson et al., 2013]
chromosome 4			
4p16.3–p15.33	3.6–11.4	mild to moderate	Basinko et al., 2008*
4p16.1–p15.2	n.a.	mild to moderate	—
4q32–q33	n.a.	mild to moderate	—
4q33–qter	n.a.	mild to moderate	—
4q33–q35.1	n.a.	mild to moderate	—
4q34	n.a.	mild to moderate	Caliebe et al., 1997
4q34.1–q34.3	n.a.	none	—
4q34.1–q34.3	n.a.	mild to moderate	—
4q35.1	180–191.3	mild to moderate	Bendavid et al., 2007

*In this reference two additional comparable cases from the literature are mentioned.

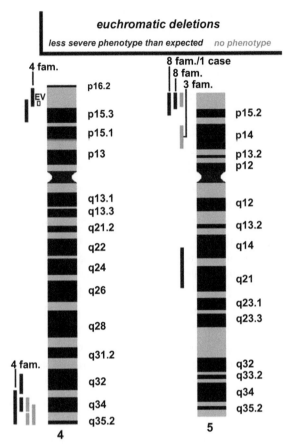

Figure 24 Euchromatic deletions with no or less severe clinical consequences than expected are shown for chromosomes 4 and 5 as described in Figure 23. A euchromatic variant region is highlighted by the abbreviation EV; where no deletion variants were reported yet is symbolized by an empty square along the corresponding region.

> Loss of copy numbers: A probe for 4q35 to 4qter (e.g., commercially available subtelomeric) can be recommended for euchromatic CG-CNVs of chromosome 4.

5.2.2.5. Chromosome 5

The following are reported as inert deletions of chromosome 5: del(5)(p14p14) (13.8 Mb, ~9.4 Mb; ~6.4 Mb), del(5)(p15.2) (~9.6 Mb), and del(5)(p15.3) (~8.1 Mb) [Barber, 2005; Barber, Brasch-Anderson et al., 2013]. Eight further families with similar del(5)(p15.3) (8.2 to 9.5 Mb) and slight phenotypic abnormalities are summarized by Barber (2005) and Barber, Brasch-Anderson et al. (2013), and a single case with slight phenotypic consequences only is reported by Rossi et al. (2005). Also Barber (2005) and Barber, Brasch-Anderson et al. (2013) collected eight families with variable outcome of Cri-du-Chat syndrome, including symptomless family members. Further one case is reported with a mild phenotype but deletion of 5q14.3 to 5q21.3 (position hg18: 83.6-104.7 Mb) [Tonk et al., 2011] (for summary see Figure 24).

Loss of copy numbers: A probe for 5p15.3 to 5pter (e.g., commercially available subtelomeric) can be recommended for euchromatic CG-CNVs of chromosome 5.

5.2.2.6. Chromosome 6
Two stretches in chromosome 6 were found in one family, each that could be deleted without clinical consequences: 6pter to 6p25 (size n.a.) and 6q22.31 to 6q23.1 (position hg18: 93.5 to 103.4 Mb [Hansson et al., 2007]) [Barber, 2005; Barber, Brasch-Anderson et al., 2013] (for summary see Figure 25).

Note: Imprinting may cause problems here (see section 5.2.1.6) [Liehr, 2010].

Loss of copy numbers: No probe can be recommended yet for euchromatic CG-CNVs of chromosome 6.

5.2.2.7. Chromosome 7
A terminal deletion del(7)(p22) ($\sim$5.5 Mb) was seen as a harmless euchromatic CG-CNV in one family so far [Barber, 2005; Barber, Brasch-Anderson et al., 2013] (for summary see Figure 25).

Note: Imprinting may cause problems here (see section 5.2.1.7) [Liehr, 2010].

Loss of copy numbers: No probe can be recommended yet for euchromatic CG-CNVs of chromosome 7.

5.2.2.8. Chromosome 8
Three families having large deletions but no clinical symptoms were reported: del(8)(p23.3$\sim$23.1) ($\sim$6.1 Mb), del(8)(p23.1) ($\sim$6.2 Mb), and del(8)(q24.13q24.22) ($\sim$4.2 Mb) [Barber, 2005; Barber, Brasch-Anderson et al., 2013]. Thus, also deletions in the EV in 8p23.1 were reported (for summary see Figure 25).

Loss of copy numbers: Probes for the EV in 8p23.1 (e.g., RP11-122N11 and RP11-24D9 [Barber, pers. communication]) can be applied for euchromatic CG-CNVs of chromosome 8.

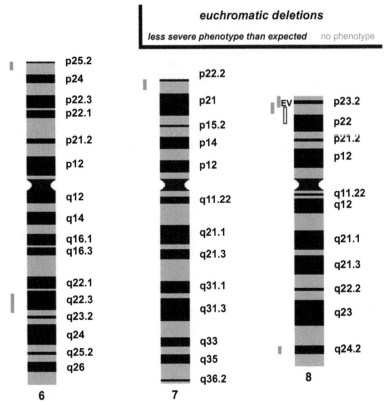

Figure 25 Euchromatic deletions for chromosomes 6 through 8 as described in Figures 23 and 24. Abbreviations: EV = euchromatic variant.

5.2.2.9. Chromosome 9

One family with asymptomatic deletion in chromosome 9 is known by presenting a del(9)(q31.2q32) ($\sim$11.4 Mb) [Barber, 2005; Barber, Brasch-Anderson et al., 2013]. The EV in 9p13.1 to 9p11 can be reduced in size as CG-CNV as demonstrated by Kosyakova et al. (2013)-see also section 5.2.19. The corresponding deletion in the EV in 9q13 to 9q21.11 has also been reported [Barber, 2005; Barber, Brasch-Anderson et al., 2013]. Finally a subtelomeric deletion in 9qter was reported to be nondeleterious [Ballif et al., 2000] (for summary see Figure 26).

> **Loss of copy numbers: The euchromatic CG-CNV characterization in chromosome 9 is restricted to centromere-near/centromere-flanking regions, which colocalize with EVs. Probes as reported in Kosyakova et al. (2013) can be applied: RP11-402N8 and/or RP11-128P23 in 9p and RP11-211E19 and/or RP11-88I18 in 9q.**

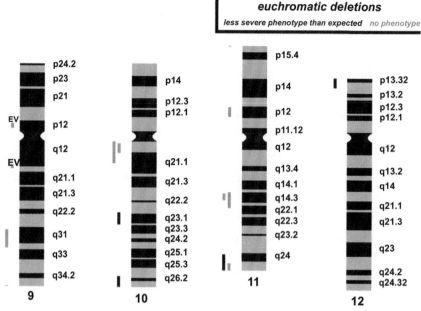

Figure 26 Euchromatic deletions for chromosomes 9 through 12 as described in Figure 23. Abbreviation: EV = euchromatic variant.

5.2.2.10. Chromosome 10

The regions 10q11.2 to 10q11.23 (position hg18: 45.4-51.6 Mb [Bisgaard et al., 2007]) and 10q11.2 to 10q21.2 (~13.3 Mb) were deleted in two clinically healthy families [Barber, 2005; Barber, Brasch-Anderson et al., 2013]. A large deletion present in 11 members of a family as del(10)(q22.3q23.31) (position hg18: 81.7-88.9 Mb [Balciuniene et al., 2007]) was associated exclusively with autism and no other abnormalities. A del(10)(q26.13) (~6.6 Mb) was combined with a dup(5)(q35qter) and was associated with dysmorphism (for summary see Figure 26).

> **Loss of copy numbers: Euchromatic CG-CNVs in chromosome 10 are restricted to the long arm of the centromere-near/centromere-flanking region short arm; pericentric probes can be found in genome browsers or elsewhere [Liehr et al., 2006].**

5.2.2.11. Chromosome 11

For chromosome 11 the following regions were reported to be lost without clinical problems for the corresponding carrier families: 11p12 to 11p12 (~6.1 Mb), 11q14.3 to 11q14.3 (~3.6 Mb), 11q14.2 to 11q22.1 (91.3-100.8 Mb [Goumy et al., 2008]), and 11q25 to 11qter (2 families; no molecular data) [Barber, 2005; Barber, Brasch-Anderson et al., 2013]. Only slight developmental delay was observed in a two-generation family with a del(11)(q24.2) (~9.6 Mb), and minor abnormalities were present in two cases with a del(11)(p15.5) (0.36 Mb) (for summary see Figure 26).

> *Note:* Imprinting may cause problems here (see section 5.2.1.11) [Liehr, 2010].

> **Loss of copy numbers: No probe can be recommended yet for euchromatic CG–CNVs of chromosome 11.**

5.2.2.12. Chromosome 12

Deletion of 12pter to 12p13.31 (position hg18: 0-6.0 Mb) [Madrigal et al., 2012] or 12p13.33 (position hg18: 0-1.6 Mb) [Baker et al., 2002] led to variable and overall minor symptoms (for summary see Figure 26).

> **Loss of copy numbers: No probe can be recommended yet for euchromatic CG–CNVs of chromosome 12.**

5.2.2.13. Chromosome 13

Deletions in 13q14 to 13q14 (~10 Mb), 13q21 to 13q21 (~16 Mb), 13q21.1 to 13q21.31 (position hg18: 53.5-64.2 Mb [Roos et al., 2008]), and 13q21.1 to 13q21.33 (position hg18: 53.6-68.1 Mb [Filges et al., 2009]) were not associated with any clinical signs in the corresponding families [Barber, 2005; Barber, Brasch-Anderson et al., 2013]. A del(13)(q14.1q21.3) (~19.9 Mb) also was associated with a normal phenotype apart from leukokoria (white pupillary reflex) and a del(13)(q31.1q31.1) (position hg18: 80.3~80.5-83.4~83.6 Mb [Bisgaard et al., 2007; Barber, 2005; Barber, Brasch-Anderson et al., 2013]) as was a del(13)(q14.3q21.2;q22.1q22.1) (hg18: 53.1-61.4 and 72.9-74.8 [Jobanputra et al., 2005]. A submicroscopic del(13)(q34) [Bedoyan et al., 2004] showed slight phenotypic signs only (for summary see Figure 27).

> **Loss of copy numbers: A probe for the euchromatic CG–CNV in 13q21.1-q21.2 should be applied.**

5.2.2.14. Chromosome 14

Deletions of band 14q12~13 (~10 Mb) or 14q31 (~8.2 Mb), respectively, only led to minor signs and symptoms in the corresponding carriers [Barber, 2005; Barber, Brasch-Anderson et al., 2013] (for summary see Figure 27).

> *Note:* Imprinting may cause problems here (see section 5.2.1.14) [Liehr, 2010].

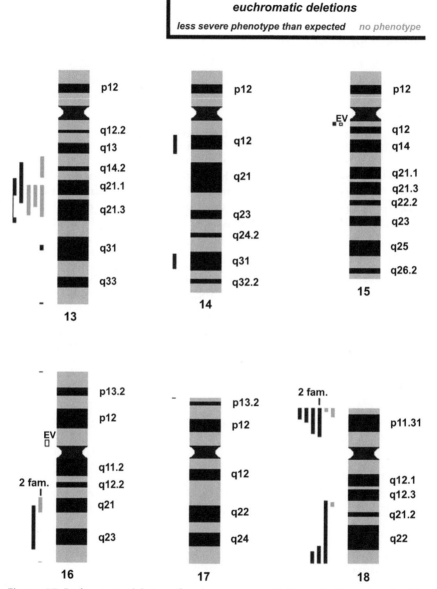

Figure 27 Euchromatic deletions for chromosomes 13 through 18 as described in Figures 23 and 24. Abbreviation: EV = euchromatic variant.

Loss of copy numbers: No probe can be recommended yet for euchromatic CG-CNVs of chromosome 14.

5.2.2.15. Chromosome 15

A deletion in 15q11 to 15q12 ($\sim$2 Mb) was accompanied by only slight mental retardation in this familial case supporting possibly a deletion in the EV located there [Barber, 2005; Barber, Brasch-Anderson et al., 2013] (for summary see Figure 27).

Note: Imprinting may cause problems here (see section 5.2.1.15) [Liehr, 2010].

Loss of copy numbers: The euchromatic CG-CNVs characterization in chromosome 15 is restricted to the centromere-near/ centromere-flanking region in the long arm, which colocalizes with an EV. Probes as reported in Liehr et al. (2006) can be applied.

5.2.2.16. Chromosome 16

Two families without any clinical signs irrespective of their deletions in 16q13 to 16q22 ($\sim$7 Mb) or 16q21 to 16q21 ($\sim$7 Mb) are known from the literature [Barber, 2005; Barber, Brasch-Anderson et al., 2013]. Also only molecular cytogenetically detectable deletions in 16pter ranging from 2.7 to 268 kb lead to almost no phenotypic effects [Horsley et al., 2001]. Callen et al. concluded already in 1993 that

> deletions involving regions more distal, from 16q22.1 to 16q24.1, were associated with relatively mild dysmorphism. One region of the long arm, q24.2 to q24.3, was not involved in any deletion; (…) presumably, monosomy for this region is lethal. In contrast, patients with deletions of 16q21 have a normal phenotype. [Callen et al., 1993]

Also the subtelomeric region in 16q is prone to loss of copy numbers [Weise et al., 2008]. To the best of our knowledge no deletion of the EV region in chromosome 16p was reported yet (for summary see Figure 27).

Loss of copy numbers: No probe can be recommended yet for euchromatic CG-CNVs of chromosome 16.

5.2.2.17. Chromosome 17

A deletion of 17pter (size 1.3 Mb) combined with a duplication of chromosome 12q24.33 (position hg18: 130.1-132.3) led to only minor symptoms [Schoumans et al., 2005] (for summary see Figure 27).

Loss of copy numbers: No probe can be recommended yet for euchromatic CG-CNVs of chromosome 17.

5.2.2.18. Chromosome 18

One family, each with a large deletion in 18p or 18q has no clinical symptoms: one with a del(18)(p11.31) (~4.4 Mb) and another one with a del(18)(q21.1q21.1) (position hg18: 42.6-44.8 Mb [Bisgaard et al., 2007]) together with a dup(4)(q35.1q35.2) [Barber, 2005; Barber, Brasch-Anderson et al., 2013]. Recently another report added a family with a small terminal deletion 18pter to 18p13.3 (~1.9 Mb in size) [Srebniak et al., 2011].

Furthermore, there are four families with minor symptoms like short stature, having del(18)(p11.2) (2 families with ~14.3 Mb), del(18)(p11.23) (7.2 Mb), or a del(18)(p11.21) (~12.9 Mb). Also solely dysmorphic features were associated in four families with del(p11.3) (~5.7 Mb) and del(18)(q2? 1) (~30.8 Mb), del(18)(q22.3) (~8.6 Mb) or del(18)(q23) (~5.7 Mb), respectively (for summary see Figure 27).

Loss of copy numbers: No probe can be recommended yet for euchromatic CG-CNVs of chromosome 18, even though in some cases commercially available subtelomeric probes can be helpful.

5.2.2.19. Chromosome 19

There is one case of a del(19)(p13) (no molecular data available) combined with a del(21)(q21.1) and no obvious clinical signs [Barber, 2005; Barber, Brasch-Anderson et al., 2013] (for summary see Figure 28).

Loss of copy numbers: No probe can be recommended yet for euchromatic CG-CNVs of chromosome 19.

5.2.2.20. Chromosome 20

Exclusively dysmorphic features were observed in two cases with partial deletions of chromosome 20p: del(20)(p12.2p11.2) (~5.8 Mb), or a del(20)(p13) (~4.4 Mb) together with a dup(6)(pterp23) [Barber, 2005; Barber, Brasch-Anderson et al., 2013] (for summary see Figure 28).

Note: Imprinting may cause problems here (see section 5.2.1.20) [Liehr, 2010].

Loss of copy numbers: No probe can be recommended yet for euchromatic CG-CNVs of chromosome 21.

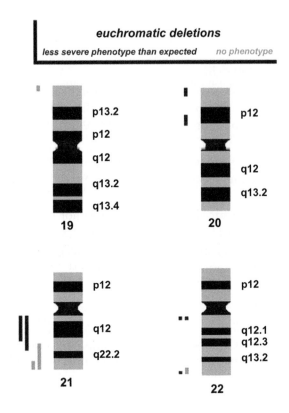

Figure 28 Euchromatic deletions for chromosomes 19 through 22 as described in Figure 23.

5.2.2.21. Chromosome 21

As mentioned in section 5.2.2.19 there is one case of a del(21)(q21.1) (no molecular data available) combined with a del(19)(p13) and no obvious clinical signs. A del(21)(q11q21.3) (~17.3 Mb) was associated only with dislocated hips, and in a del(21)(pterq21.2) (~21.6 Mb) only minor clinical signs were seen [Barber, 2005; Barber, Brasch-Anderson et al., 2013]. Also a similar case with a terminal deletion of 21q22.3 of 3.4 Mb size was not associated with gross aberrations; in this case reported comparable cases without clinical signs are summarized as well [Bertini et al., 2008] (for summary see Figure 28).

> **Loss of copy numbers:** No probe can be recommended yet for euchromatic CG-CNVs of chromosome 21, even though in some cases commercially available subtelomeric probes can be helpful.

5.2.2.22. Chromosome 22

There is one case of a del(22)(pterq11.21) ($\sim$4.1 Mb) and no clinical signs [Barber, 2005; Barber, Brasch-Anderson et al., 2013]. Dysmorphisms and/or heart defects were the only effects of a del(22)(q11.2q11.2) ($\sim$2 Mb), del(22)(q13.3q13.3) (no molecular data available), or a del(22)(pterq11.2) ($\sim$12.4 Mb) together with a del(4)(q35.2) [Barber, 2005; Barber, Brasch-Anderson et al., 2013] (for summary see Figure 28).

> **Loss of copy numbers: No probe can be recommended yet for euchromatic CG-CNVs of chromosome 22, even though in some cases commercially available subtelomeric probes or centromere-near ones [Liehr et al., 2006] can be helpful.**

5.2.2.23. X-Chromosome

Basically everything necessary concerning CG-CNV of the X-chromosome was stated in section 5.2.1.23. Here it may be added that the loss of an X-chromosome is the only known viable monosomy in human, however, being connected with Turner syndrome; also mosaicism is suggested to play a role here.

> **Loss of copy numbers: Any probe determining the copy number of the X-chromosome in meta- or interphase can be applied.**

5.2.2.24. Y-Chromosome

Loss of the whole Y-chromosome leads to monosomy X (i.e., Turner syndrome; see section 5.2.2.23). Loss as gain (see section 5.2.1.24) of any part of the Y-chromosome is viable, but according to the lost region, may be connected with a female habitus and/or infertility.

> **Loss of copy numbers: Any probe determining the copy number of the Y-chromosome in meta- or interphase can be applied.**

5.2.3. X-Autosome Translocations

In female carriers of unbalanced X-autosome translocations, skewed X-chromosome inactivation may lead to much less severe clinical outcome than expected. This was the case, for example, for a der(X)t(X;12)(q28;12p11.21) [Dufke et al., 2006], a der(X)t(X;18)(q27;q22) [Fusco et al., 2011], a der(X)t(X;1)(q28;q32.1) [Guo et al., 2010], or a der(X)t(X;15)(q22.3;q11.2) [Stankiewicz et al., 2006]. Thus, CG-CNVs involving even large euchromatic stretches of any other chromosome may be viable or (almost) without clinical symptoms.

Any probe determining the presence of X-chromosome parts (best suited a whole chromosome paint) and a commercially available probe for the XIST region for example, is helpful here.

5.3. SUBMICROSCOPIC CNVs (MG-CNVs)

MG-CNVs (Chapter 2, section 2.8) are discussed briefly in this third and last part of Chapter 5. Concerning MG-CNVs the best resource for their analysis may be the Database of Genomic Variants (http://projects.tcag.ca/variation/ and http://dgvbeta.tcag.ca/gb2/gbrowse/dgv2_hg18/). This specific web page provides the impression that practically every part of the human genome may be subjected to gain or loss of copy numbers, segmental duplications, small inversions, or submicroscopic complex rearrangements. Also MG-CNVs may colocalize with CG-CNVs (see the appendix).

The multitude of available online resources for genomic structural variation analysis was summarized recently by Sneddon and Church (2012) (see Chapter 7).

5.3.1. Benign Variants

Tens of thousands of MG-CNVs are collected in the Database of Genomic Variants (http://projects.tcag.ca/variation/), thus it is impossible to include all those variants in a book. To analyze an aCGH result the Database of Genomic Variants and many other online resources (see Chapter 7) have to be consulted, as those are regularly updated and can be specifically checked for the corresponding region(s) of interest.

In addition, there are chromosome-specific review articles available; for example, by 2008 Jobling reviewed all known MG-CNVs for the Y-chromosome. This data is also included in the mentioned databases. The most frequently reported MG-CNVs were reviewed by Weise and co-workers (2008). Those are included in the chromosome-specific figures in the appendix, summarizing all CG-CNVs covered in this book.

5.3.2. Pathologic Variants

For pathological variants the UCSC (University of California, Santa Cruz) genome browser (http://genome.ucsc.edu/) linked to the Centre for the Development and Evaluation of Complex Interventions for Public Health Improvement (DECIPHer) project (https://decipher.sanger.ac.uk) and the International Standards for Cytogenomic Arrays (ISCA) Consortium (https://www.iscaconsortium.org/) seems to be the best-suited resource [Swaminathan et al., 2012]. In general, pathological MG-CNVs should overlap with critical regions of microdeletion and microduplication (MDD) syndromes [Weise et al., 2012]. Interestingly, all MDD syndromes form based on the same mechanism that was initially detected in Charcot-Marie-Tooth disease Type 1A (CMT1A; microduplication in 17p12) and its counterpart Hereditary Neuropathy with Liability to Pressure Palsies (HNPP—microdeletion in 17p12) [Roa and Lupski, 1994]. The principle is simple: one or more dosage gene(s) are flanked by repetitive homologous sequences (REP elements), which are not more distant than a few megabasepairs. It is now just a matter of statistics that an unequal crossing over of these REP elements happens, leading to gametes with a microdeletion and a corresponding microduplication. If one of these gametes is fertilizing another normal gamete there is a newborn with MDD syndrome (Figure 29).

As for CG-CNVs there are many MG-CNVs that are not pathologic in a parent, but are present in the clinically affected offspring (e.g., [Cuoco et al., 2011]). Two explanations are available for microduplication syndromes (Figure 30) and several theories have been issued to explain this for microdeletion syndromes (see Chapter 1, section 1.4 and Figure 31).

Apart from the expected situation of normal parents and microduplication in the affected child (Figure 30A), the identical

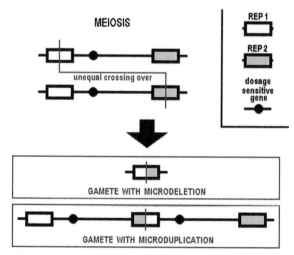

Figure 29 Schematic depiction for the formation of MDD syndromes.

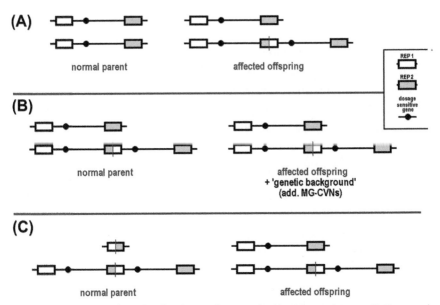

Figure 30 Situations found in families with microduplication syndromes. **A.** Expected situation: Both clinically normal parents have no copy number changes in the disease-causing critical region, but the offspring has a microduplication. **B.** Unexpected situation 1: One parent and the offspring show the identical microduplication; the two-hit model explains this by the presence of otherwise unproblematic additional MG-CNVs. **C.** Unexpected situation 2: One parent has a microduplication and a microdeletion, thus in summary a balanced situation only detectable by FISH, and provides a gamete with a microduplication to its offspring.

microduplication may be present in one parent and the offspring. The latter is explained by another genetic background in the patient than in the healthy person. Girirajan et al. (2010) introduced here more concretely the two-hit model-that is, that a disease phenotype can be observed only if additional imbalances (MG-CNVs) are present in the genome of the patient (Figure 30B). Also possible is that one parent appears balanced in aCGH, suggesting a situation like that shown in Figure 30A, but in fact has a microduplication on one and a microdeletion on the other of the homologous chromosomes (Figure 30C).

Aside from what was discussed to explain aforementioned discrepancies of parents and offspring for microduplication syndromes, for microdeletion syndrome one more rationalization has been identified so far. Figure 31 shows first the normal situation (Figure 31A), followed by the genetic background/two-hit model (Figure 31B) and the microduplication present

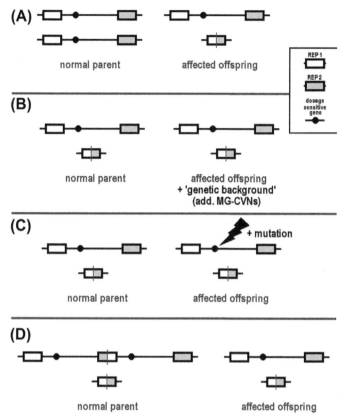

Figure 31 Situations found in families with microdeletion syndromes. **A.** Expected situation: Both clinically normal parents have no copy number changes in the disease-causing critical region, but the offspring has a microdeletion. **B.** Unexpected situation 1: One parent and the offspring show the identical microdeletion; the two-hit model explains this by the presence of otherwise unproblematic additional MG-CVNs. **C.** Unexpected situation 1a: One parent and the offspring show the identical microdeletion; an otherwise recessive mutation in the present gene copy promotes the disease. **D.** Unexpected situation 2: One parent has a microduplication and a microdeletion, thus in summary a balanced situation only detectable by FISH, and provides a gamete with a microdeletion to its offspring.

with a microdeletion in a parent (Figure 31D). Recessive mutations in the monosomic region were already proven as disease causing (Figure 31C; see Chapter 1, section 1.4).

CHAPTER 6

CG-CNVs in Genetic Diagnostics and Counseling

6.1. CG-CNVs IN DIAGNOSTICS

CG-CNVs may be identified, studied, and characterized by many different approaches. Normally, CG-CNVs are found in banding cytogenetics (section 6.1.1) and/or molecular cytogenetics (section 6.1.2), and molecular approaches are restricted to the detection of euchromatic CG-CNVs (section 6.1.3). The main problem with CG-CNVs is distinguishing them from meaningful cytogenetic imbalances (section 6.1.4).

6.1.1. First Steps: Banding Cytogenetics

Even though banding cytogenetics has been considered "dead" for over 40 years, it is still the most often used approach worldwide to study the human genome. Thus, it is the major method by which CG-CNVs usually are detected. Banding cytogenetics, most often GTG-banding or R-banding, is a standard approach in each cytogenetic laboratory. After a heterochromatic CG-CNV has been spotted, CBG and NOR staining are performed as well. Principles, methods, and possible problems for these basic cytogenetic approaches are summarized elsewhere [Liehr, 2009]. Typical results of GTG-, CBG-, and NOR-stained CG-CNVs were depicted in Figure 1 (in Chapter 1).

In the 1980s and 1990s DAPI/DA banding also was applied to further characterize the heterochromatic CG-CNVs [Babu and Verma, 1986]; however, a lack of specificity was proven for this approach [Lin et al., 1990, Verma et al., 1991]. Still, in DAPI staining applied in molecular cytogenetics (see section 6.1.2), heterochromatic CG-CNVs may appear better or more expressed than in GTG banding. Thus, even CG-CNVs may be identified after a FISH test as unexpected spillover.

6.1.1.1. Other Cytogenetic Approaches

Apart from GTG, CBG, NOR, or fluorescence staining, there might be more suited cytogenetic approaches to determine and subclassify

Benign and Pathological Chromosomal Imbalances
ISBN 978-0-12-404631-3

acrocentric short-arm heteromorphisms, but nothing else apart from GTG- and C-banding as well as NOR-staining has been accepted into routine diagnostics. CG-CNV variants of acrocentric short arms and centromeric regions of all chromosomes were reported based on staining with mithramycin [Wachtler and Musil, 1980] or quinacrine mustard [Wachtler and Musil, 1980], an indirect immunoperoxidase technique to detect 5-methylcytosine (5MeC)-rich DNA [Okamoto et al., 1981], high-resolution dual Q-R banding [Kamei et al., 1986], CDG combined with oblique epi-illumination [Wahedi and Pawlowitzki, 1987], by in situ treatments with restriction endonucleases AluI, NdeII, and Sau3AI followed by Giemsa staining [De Cabo et al., 1991], or recently by staining pericentromeric heterochromatin regions using acridine orange [Kuznetzova et al., 2012]. Interestingly, an influence of the studied cell type has been shown for the expression of CG-CNVs [Fernández et al.; 1995].

6.1.2. Molecular Cytogenetics

If cytogenetics has picked up a potential CG-CNV, its further molecular cytogenetic characterization is driven by one main question: Is what we see a CG-CNV without clinical meaning, or is it a possible chromosomal rearrangement leading to a meaningful euchromatic imbalance? In Chapter 5, corresponding FISH probes are mentioned at the appropriate places, and typical FISH results are shown in the four Color Plates of this book. Comprehensive descriptions of current FISH approaches can be found elsewhere [Liehr, 2009].

6.1.3. Molecular Genetics

As already mentioned, molecular genetics in general is not suited to characterize heterochromatic CG-CNVs in more detail. However, molecular genetics using the aCGH approach can easily detect large chromosomal imbalances caused by euchromatic CG-CNVs. (Technical issues of aCGH are covered elsewhere [Liehr, 2009].) Thus on the one hand, aCGH may be helpful in characterizing harmless UBCAs and EVs and distinguishing them from harmful ones, but aCGH is not necessarily the first choice to characterize mosaic sSMC in more detail. aCGH in routine diagnostics often fails to detect sSMC present in less than 50% of the cells [Liehr, 2012].

6.1.4. How to Characterize CG-CNVs and MG-CNVs in Diagnostics

6.1.4.1. Heterochromatic CG-CNVs

Heterochromatic CG-CNVs discovered in banding cytogenetics should be characterized according to the following scheme:

1. Apply CBG staining.
2. Apply NOR staining if appropriate.
3. Do parental studies to track the origin of the potential heterochromatic CG-CNV.
4. If one of the parents has the identical CG-CNV make sure it is not part of a balanced rearrangement, potentially not inherited in a balanced way to the index patient.
5. If points 3 and 4 could not be resolved convincingly apply suited FISH probes in a molecular cytogenetic setting; for corresponding probes see Chapter 5, section 5.1.

The overall goal must be to rule out the possibility the CG-CNV is no heteromorphism, but a meaningful rearrangement leading to a disease-causing phenotype. This is valid for all diagnostic cases, but most importantly in prenatals.

6.1.4.2. Euchromatic CG-CNVs

The exact characterization of euchromatic CG-CNVs is more sophisticated and dependant in the methodologies and probe sets the individual laboratory has available. A euchromatic CG-CNV looks like any other chromosomal rearrangement leading to gain or loss of copy numbers.

However, if regions highlighted in Figure 5 as EVs are involved, or the pericentric regions, the cytogeneticist needs to be alerted. Unless the laboratory has EV-specific FISH probes in stock it should send the case to accordingly specialized institutions.

A scheme for the characterization of euchromatic CG-CNVs is suggested here:

1. Apply CBG or NOR staining if appropriate.
2. Do parental studies to track the origin of the potential euchromatic CG-CNV.
3. Compare the detected imbalance with data from the literature (Chapter 5, section 5.2, and the appendix).
4. If one parent has the identical CG-CNV make sure it is not part of a balanced rearrangement, potentially not inherited in a balanced way to the index patient.

5. If points 3 and 4 could not be resolved convincingly or if an EV might be involved, apply suited FISH probes in a molecular cytogenetic setting; for corresponding probes see Chapter 5, section 5.2. Also aCGH may be applied to narrow down the exact breakpoints of the region involved in the imbalance caused by the CG-CNV.

6.1.4.3. MG-CNVs

Identification and characterization of MG CNVs became possible only after introduction of aCGH in research and diagnostics. Accordingly, aCGH is the method of choice to detect submicroscopic chromosomal imbalances. It is standard to verify detected gains or loss of copy numbers by a second independent approach (in most cases FISH or MLPA techniques) in aCGH. The main part of the work in aCGH, however, is not the tracking of the MG-CNVs, but the interpretation of which is meaningful and which is not, based on a computer-assisted search of multiple databases (see Chapter 7).

6.2. CG-CNVs AND MG-CNVs IN REPORTING AND GENETIC COUNSELING

This book is not intended to review all the ethical issues, legal problems, and quality issues correlated with genetic counseling—these considerations may be found elsewhere [Kristoffersson, 2008; Kristoffersson et al., 2010].

6.2.1. Heterochromatic CG-CNVs in Reporting and Genetic Counseling

For heterochromatic CG-CNVs there is an ongoing discussion regarding which kind of heteromorphism should be reported and which not. The consensus is to report inversions, fragile sites, and sSMCs, but according to Brothman et al. (2006), other heteromorphic CG-CNVs are not meant to be reported. Also Gardner et al. (2012) states: "Obviously enough, it is crucial that the cytogeneticist distinguishes normal variant from abnormality. Generally, there is no point in reporting a particular variant to the referring practitioner or the patient". Also in an earlier edition of the same textbook, Gardner and Sutherland (2004) stress the problem that "the counselor may thoroughly understand the presumed harmlessness of a variant chromosome, but the person in whose family it has been discovered may react" in an inappropriate way with "the worst possible response for a couple to choose to terminate a pregnancy because of an overinterpreted variant chromosome."

Considering all these arguments, nonetheless I think that heterochromatic CG-CNVs as defined in this textbook need to be reported. The main reason is the following: No cytogeneticist would refrain from reporting an enlargement of an acrocentric short arm of the size of a chromosome 16, or a similar large block of 1q12, 9q12, 16q11.2, or Yq12. Every laboratory would mention this kind of gross heterochromatic CG-CNV in its report, even though the previous arguments against reporting are the same as for less extended heterochromatic CG-CNVs. As this fact is indubitable this means trying to avoid the reporting of heterochromatic CG-CNVs is not always possible. Thus, one laboratory will tend to report a CG-CNV whereas another will not report the identical one, maybe in the same patient. Thus, the patient or family is even more confused, wondering why laboratory A found an alteration at chromosome 9 but laboratory B did not.

Finally, what do we do about CG-CNVs detected in a prenatal setting in the child but not in the mother or the putative father? This needs to be reported, especially if the corresponding laboratory has no molecular methods for the clarification if it is a CG-CNV or a meaningful rearrangement or a hint on another paternity.

A compromise may be to clearly state in the report to the clinician that overall a normal result was obtained. Nonetheless, a karyotype should be provided that includes the detected heteromorphism.

In general, the task of a counselor is to make difficult context and interrelationships understandable to the patient and family. Thus, explaining the meaning of a heterochromatic CG-CNV is a normal part of the highly responsible role of a counselor in the scenario of genetic counseling. A counselor has to find the balance between providing all available information and presenting it in a way that really reaches the counseled persons.

Some national human genetic societies recommend that genetic counseling is done before each genetic analysis; this has even been regulated by law recently (e.g., in Germany) [Liehr, 2012]. Thus, the possibility that a CG-CNV may be found is discussed before getting the results. The same should be done as well for aCGH before testing.

6.2.2. Euchromatic CG-CNVs and MG-CNVs in Reporting and Genetic Counseling

Euchromatic CG-CNVs and MG-CNVs have in common that they might contain genetically relevant material. The impact for the carrier is more or less the same, since in both cases deleterious gain or loss of copy numbers has to be distinguished from nondeleterious ones.

If identical euchromatic CG-CNVs and MG-CNVs are found either in one of the unaffected parents, or in independent normal controls, they most probably have no direct phenotypic consequences; however, low penetrance and variable expressivity of the phenotype may complicate the analysis and genetic counseling (see also Figures 29–31). While euchromatic CG-CNVs are relatively scarce and summarized in this textbook, the publicly available MG-CNV databases provide many datasets that assist in making decisions about the clinical significance of imbalances detected by microarrays. Examples of such databases are summarized in Chapter 7.

When determined as de novo in origin genomic imbalances are considered more likely pathological [Tyson et al., 2005]. This can be further supported if the implicated region contains gene(s) with functions compatible with the abnormal clinical findings or previously described patients with a similar genomic imbalance and a similar phenotype. The de novo occurrence of copy number alteration, however, is not absolute evidence of its pathogenicity and caution must be exercised for possible nonpaternity. Moreover genetic modifiers or thresholds involving other copy-number alterations could play a role in the manifestation of clinical features, or other independent mutations elsewhere in the genome may obfuscate the interpretation of such data.

In contrast to heterochromatic CG-CNVs the reporting of any kind of euchromatic CG-CNVs is undisputed. Counseling nonetheless is no easier since information on euchromatic CG-CNVs is still uncommon. What makes euchromatic CG-CNVs particularly difficult for counseling is the unpredictability of the clinical outcomes for individuals of the same family. However, there is no other choice but to discuss the results in an open way with the people concerned.

In the early days of aCGH all MG-CNVs were reported; now the tendency is to report only potentially meaningful copy number alterations. Still, most aCGH reports at least mention that additional MG-CNVs were found that are considered heteromorphisms, and counselors can get this information on request. Counseling of aCGH results did not make the life of the specialists involved in this task easier. Because the methodology—at least roughly—needs to be explained to the patients, it is an easy task to include the information on the potentially meaningless MG-CNVs detectable here. As mentioned in section 6.2.1, at this point of genetic counseling it would be easy to state that such MG-CNVs also can reach the microscopic level and then appear as CG-CNVs.

Online Resources

More and more online resources from the World Wide Web have become available. Because some are subject to quick change, they may no longer be available or may have moved since they were first accessed. Resources available as of spring 2013 are listed in this chapter.

7.1. CG-CNVs

These resources are directly related to CG-CNVs:
- Chromosome Anomaly Collection (John Barber)
    http://www.ngrl.org.uk/Wessex/collection/index.htm
- Small supernumerary marker chromosomes (Thomas Liehr)
    http://www.fish.uniklinikum-jena.de/sSMC.html

Other online resources about chromosome abnormalities and variants:
- Borgaonkar Online Database
    http://www.wiley.com/legacy/products/subject/life/borgaonkar/access.html
- Centre for the Development and Evaluation of Complex Interventions for Public Health Improvement (DECIPHer) project
    https://decipher.sanger.ac.uk
- Cytogenetic Data Analysis System (CyDAS)
    http://www.cydas.org/
- European Cytogenetic Association Register of Unbalanced Chromosome Abnormalities (ECARUCA)
    www.ecaruca.net/

Online resources about molecular cytogenetics:
- mFISH: Basics and literature on multicolor fluorescence in situ hybridization application
    http://www.fish.uniklinikum-jena.de/mFISH.html
- Resources for Molecular Cytogenetics
    http://www.biologia.uniba.it/rmc/
- SKY/M-FISH and CGH database
    http://www.ncbi.nlm.nih.gov/sky/
- Tavi's multicolor FISH page
    http://medicine.yale.edu/labs/henegariu/www/tavi/FISH.html

Benign and Pathological Chromosomal Imbalances
ISBN 978-0-12-404631-3

7.2. MG-CNVs

These resources are directly related to MG-CNVs:
- 1000 Genomes: A Deep Catalog of Human Genetic Variation
  http://www.1000genomes.org/
- Overview of structural variation
  http://www.ncbi.nlm.nih.gov/dbvar/content/overview/

Genome browsers:
- Ensembl
  www.ensembl.org/
- NCBI (National Center for Biotechnology Information)
  http://www.ncbi.nlm.nih.gov/mapview/maps.cgi?
  ORG=hum&MAPS=ideogr,est,loc&LINKS=ON&VERBOSE=
  ON&CHR=1
- UCSC (University of California, Santa Cruz)
  http://genome.ucsc.edu/

Further analyzing support web sites for variants in general:
- Copy Number Variation (CNV) Project
  http://www.sanger.ac.uk/research/areas/humangenetics/cnv/
- Copy number variation at CHOP
  http://www.research.chop.edu
- Database of genomic structural variation (dbVar)
  http://www.ncbi.nlm.nih.gov/dbvar/
- Database of genomic variants
  http://projects.tcag.ca/variation/
- Database of genomic variants archive (DGVa)
  http://www.ebi.ac.uk/dgva/
- Diploid human genome browser (HuRef)
  huref.jcvi.org/
- Human Genome Structural Variation project
  http://humanparalogy.gs.washington.edu/structuralvariation/
- International Standards for Cytogenomic Arrays (ISCA) Consortium
  https://www.iscaconsortium.org/
- ISCA Dosage sensitivity map
  http://www.ncbi.nlm.nih.gov/projects/dbvar/ISCA/index.shtml
- Segmental duplication database
  http://humanparalogy.gs.washington.edu/build36/build36.htm

Further analyzing support web sites for variants in autism:
- Autism chromosome rearrangement database
  http://projects.tcag.ca/autism/
- Autism CNV database
  http://projects.tcag.ca/autism_500k/
- Autism genetic database
  http://wren.bcf.ku.edu/

Further analyzing support web sites for variants in cancer:
- The Cancer Genome Atlas (TCGA)
  http://cancergenome.nih.gov/
- NCI recurrent aberrations in cancer database
  http://cgap.nci.nih.gov/Chromosomes/RecurrentAberrations
- Progenetix: Visualizing cancer genomics
  progenetix.tumblr.com/

Online resources about molecular genetics:
- HUSAR Bioinformatics Lab
  http://genome.dkfz-heidelberg.de/
- Microarray genomics, Princeton University
  http://www.princeton.edu/genomics/microarray/

Summary of CG-CNVs by Chromosome

Here the CG-CNVs mentioned in Chapter 5 are summarized. On the left side of each idiogram the regions tolerated as CG-CNVs (deleted or duplicated) are indicated. Also the most frequently occurring MG-CNVs are plotted in a separate column on the far left side of the scheme. The following 24 sections are to be considered carefully; they summarize in parts just single case reports. The thinner the line that marks a specific chromosomal region, the less reliable is the data; that is, unknown unique events might have contributed to the fact that the individual with the corresponding deletion or duplication was healthy or only (comparatively) minimally affected. In cases where the line is maximally broad, more than one case or family was reported having the corresponding stretch of imbalance. The heterochromatic regions are marked by two parallel lines, and these chromosomal parts may be lost or duplicated/amplified without clinical impact. Overall, this summarizing data can only be interpreted together with the underlying information, especially from Chapter 5, section 5.2.

A.1 CHROMOSOME 1

According to GRCh37/hg19 chromosome 1 has a size of 249.3 Mb. Figure 32 summarizes all known regions that have been reported at least once to be non- or only minimally deleterious for the corresponding carriers. More detailed information is presented in Figures 17 and 23.

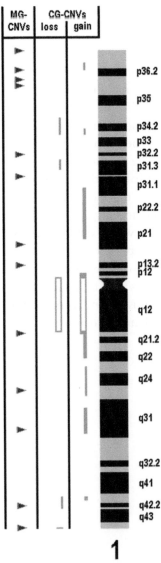

Figure 32 Idiogram of chromosome 1 according to Kosyakova et al. (2009) showing regions that, if deleted or duplicated, have only minor clinical impact. The structure of the figure is explained in the first paragraph of this appendix.

A.2 CHROMOSOME 2

According to GRCh37/hg19 chromosome 2 has a size of 243.2 Mb. Figure 33 summarizes all known regions that have been reported at least once to be non- or only minimally deleterious for the corresponding carriers. More detailed information is presented in Figures 17 and 23.

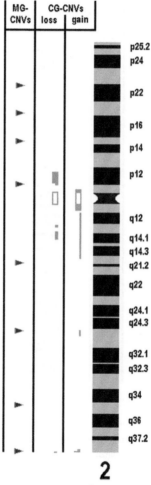

Figure 33 Idiogram of chromosome 2 according to Kosyakova et al. (2009) showing regions that, if deleted or duplicated, have only minor clinical impact. The structure of the figure is explained in the first paragraph of this appendix.

A.3 CHROMOSOME 3

According to GRCh37/hg19 chromosome 3 has a size of 198.0 Mb.
Figure 34 summarizes all known regions that have been reported at least
once to be non- or only minimally deleterious for the corresponding
carriers. More detailed information is presented in Figures 17 and 23.

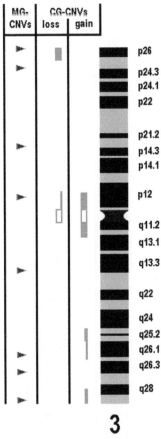

Figure 34 Idiogram of chromosome 3 according to Kosyakova et al. (2009) showing
regions that, if deleted or duplicated, have only minor clinical impact. The structure of
the figure is explained in the first paragraph of this appendix.

A.4 CHROMOSOME 4

According to GRCh37/hg19 chromosome 4 has a size of 191.2 Mb. Figure 35 summarizes all known regions that have been reported at least once to be non- or only minimally deleterious for the corresponding carriers. More detailed information is presented in Figures 18 and 24.

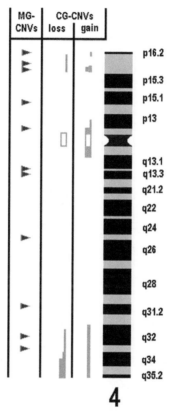

Figure 35 Idiogram of chromosome 4 according to Kosyakova et al. (2009) showing regions that, if deleted or duplicated, have only minor clinical impact. The structure of the figure is explained in the first paragraph of this appendix.

A.5 CHROMOSOME 5

According to GRCh37/hg19 chromosome 5 has a size of 181.0 Mb. Figure 36 summarizes all known regions that have been reported at least once to be non- or only minimally deleterious for the corresponding carriers. More detailed information is presented in Figures 18 and 24.

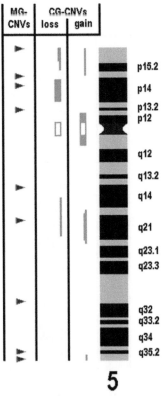

Figure 36 Idiogram of chromosome 5 according to Kosyakova et al. (2009) showing regions that, if deleted or duplicated, have only minor clinical impact. The structure of the figure is explained in the first paragraph of this appendix.

A.6 CHROMOSOME 6

According to GRCh37/hg19 chromosome 6 has a size of 171.1 Mb. Figure 37 summarizes all known regions that have been reported at least once to be non- or only minimally deleterious for the corresponding carriers. More detailed information is presented in Figures 19 and 25.

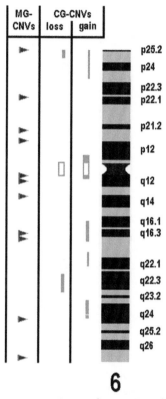

Figure 37 Idiogram of chromosome 6 according to Kosyakova et al. (2009) showing regions that, if deleted or duplicated, have only minor clinical impact. The structure of the figure is explained in the first paragraph of this appendix.

A.7 CHROMOSOME 7

According to GRCh37/hg19 chromosome 7 has a size of 159.1 Mb. Figure 38 summarizes all known regions that have been reported at least once to be non- or only minimally deleterious for the corresponding carriers. More detailed information is presented in Figures 19 and 25.

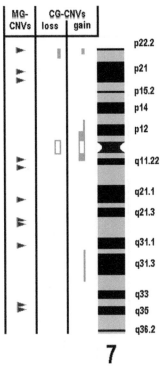

Figure 38 Idiogram of chromosome 7 according to Kosyakova et al. (2009) showing regions that, if deleted or duplicated, have only minor clinical impact. The structure of the figure is explained in the first paragraph of this appendix.

A.8 CHROMOSOME 8

According to GRCh37/hg19 chromosome 8 has a size of 146.4 Mb. Figure 39 summarizes all known regions that have been reported at least once to be non- or only minimally deleterious for the corresponding carriers. More detailed information is presented in Figures 19 and 25.

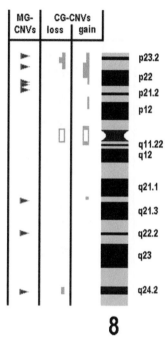

Figure 39 Idiogram of chromosome 8 according to Kosyakova et al. (2009) showing regions that, if deleted or duplicated, have only minor clinical impact. The structure of the figure is explained in the first paragraph of this appendix.

A.9 CHROMOSOME 9

According to GRCh37/hg19 chromosome 9 has a size of 141.2 Mb. Figure 40 summarizes all known regions that have been reported at least once to be non- or only minimally deleterious for the corresponding carriers. More detailed information is presented in Figures 20 and 26.

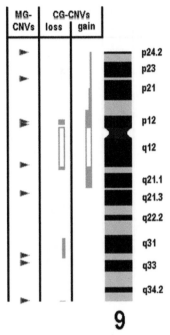

Figure 40 Idiogram of chromosome 9 according to Kosyakova et al. (2009) showing regions that, if deleted or duplicated, have only minor clinical impact. The structure of the figure is explained in the first paragraph of this appendix.

A.10 CHROMOSOME 10

According to GRCh37/hg19 chromosome 10 has a size of 135.5 Mb.
Figure 41 summarizes all known regions that have been reported at least
once to be non- or only minimally deleterious for the corresponding carriers. More detailed information is presented in Figures 20 and 26.

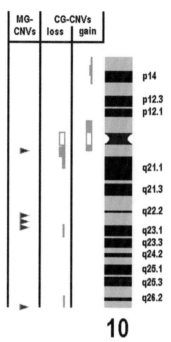

Figure 41 Idiogram of chromosome 10 according to Kosyakova et al. (2009) showing regions that, if deleted or duplicated, have only minor clinical impact. The structure of the figure is explained in the first paragraph of this appendix.

A.11 CHROMOSOME 11

According to GRCh37/hg19 chromosome 11 has a size of 135.0 Mb. Figure 42 summarizes all known regions that have been reported at least once to be non- or only minimally deleterious for the corresponding carriers. More detailed information is presented in Figures 20 and 26.

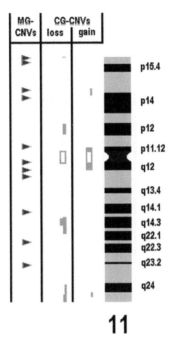

Figure 42 Idiogram of chromosome 11 according to Kosyakova et al. (2009) showing regions that, if deleted or duplicated, have only minor clinical impact. The structure of the figure is explained in the first paragraph of this appendix.

A.12 CHROMOSOME 12

According to GRCh37/hg19 chromosome 12 has a size of 133.9 Mb. Figure 43 summarizes all known regions that have been reported at least once to be non- or only minimally deleterious for the corresponding carriers. More detailed information is presented in Figures 20 and 26.

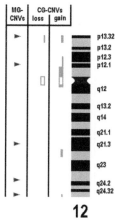

Figure 43 Idiogram of chromosome 12 according to Kosyakova et al. (2009) showing regions that, if deleted or duplicated, have only minor clinical impact. The structure of the figure is explained in the first paragraph of this appendix.

A.13 CHROMOSOME 13

According to GRCh37/hg19 chromosome 13 has a size of 115.2 Mb. Figure 44 summarizes all known regions that have been reported at least once to be non- or only minimally deleterious for the corresponding carriers. More detailed information is presented in Figures 21 and 27.

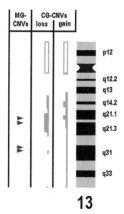

Figure 44 Idiogram of chromosome 13 according to Kosyakova et al. (2009) showing regions that, if deleted or duplicated, have only minor clinical impact. The structure of the figure is explained in the first paragraph of this appendix.

A.14 CHROMOSOME 14

According to GRCh37/hg19 chromosome 14 has a size of 107.3 Mb. Figure 45 summarizes all known regions that have been reported at least once to be non- or only minimally deleterious for the corresponding carriers. More detailed information is presented in Figures 21 and 27.

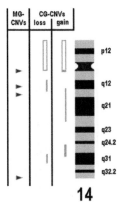

Figure 45 Idiogram of chromosome 14 according to Kosyakova et al. (2009) showing regions that, if deleted or duplicated, have only minor clinical impact. The structure of the figure is explained in the first paragraph of this appendix.

A.15 CHROMOSOME 15

According to GRCh37/hg19 chromosome 15 has a size of 102.5 Mb. Figure 46 summarizes all known regions that have been reported at least once to be non- or only minimally deleterious for the corresponding carriers. More detailed information is presented in Figures 21 and 27.

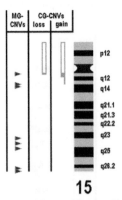

Figure 46 Idiogram of chromosome 15 according to Kosyakova et al. (2009) showing regions that, if deleted or duplicated, have only minor clinical impact. The structure of the figure is explained in the first paragraph of this appendix.

A.16 CHROMOSOME 16

According to GRCh37/hg19 chromosome 16 has a size of 90.4 Mb. Figure 47 summarizes all known regions that have been reported at least once to be non- or only minimally deleterious for the corresponding carriers. More detailed information is presented in Figures 21 and 27.

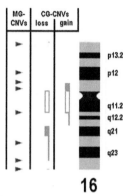

Figure 47 Idiogram of chromosome 16 according to Kosyakova et al. (2009) showing regions that, if deleted or duplicated, have only minor clinical impact. The structure of the figure is explained in the first paragraph of this appendix.

A.17 CHROMOSOME 17

According to GRCh37/hg19 chromosome 17 has a size of 81.2 Mb. Figure 48 summarizes all known regions that have been reported at least once to be non- or only minimally deleterious for the corresponding carriers. More detailed information is presented in Figures 21 and 27.

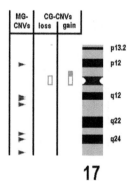

Figure 48 Idiogram of chromosome 17 according to Kosyakova et al. (2009) showing regions that, if deleted or duplicated, have only minor clinical impact. The structure of the figure is explained in the first paragraph of this appendix.

A.18 CHROMOSOME 18

According to GRCh37/hg19 chromosome 18 has a size of 78.1 Mb. Figure 49 summarizes all known regions that have been reported at least once to be non- or only minimally deleterious for the corresponding carriers. More detailed information is presented in Figures 21 and 27.

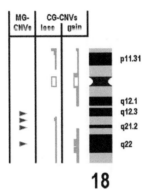

Figure 49 Idiogram of chromosome 18 according to Kosyakova et al. (2009) showing regions that, if deleted or duplicated, have only minor clinical impact. The structure of the figure is explained in the first paragraph of this appendix.

A.19 CHROMOSOME 19

According to GRCh37/hg19 chromosome 19 has a size of 59.1 Mb. Figure 50 summarizes all known regions that have been reported at least once to be non- or only minimally deleterious for the corresponding carriers. More detailed information is presented in Figures 22 and 28.

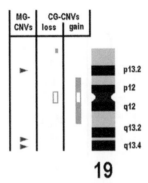

Figure 50 Idiogram of chromosome 19 according to Kosyakova et al. (2009) showing regions that, if deleted or duplicated, have only minor clinical impact. The structure of the figure is explained in the first paragraph of this appendix.

A.20 CHROMOSOME 20

According to GRCh37/hg19 chromosome 20 has a size of 63.0 Mb. Figure 51 summarizes all known regions that have been reported at least once to be non- or only minimally deleterious for the corresponding carriers. More detailed information is presented in Figures 22 and 28.

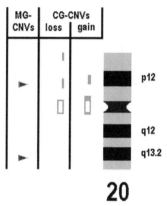

Figure 51 Idiogram of chromosome 20 according to Kosyakova et al. (2009) showing regions that, if deleted or duplicated, have only minor clinical impact. The structure of the figure is explained in the first paragraph of this appendix.

A.21 CHROMOSOME 21

According to GRCh37/hg19 chromosome 21 has a size of 48.1 Mb. Figure 52 summarizes all known regions that have been reported at least once to be non- or only minimally deleterious for the corresponding carriers. More detailed information is presented in Figures 22 and 28.

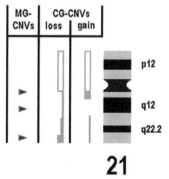

Figure 52 Idiogram of chromosome 21 according to Kosyakova et al. (2009) showing regions that, if deleted or duplicated, have only minor clinical impact. The structure of the figure is explained in the first paragraph of this appendix.

A.22 CHROMOSOME 22

According to GRCh37/hg19 chromosome 22 has a size of 51.3 Mb. Figure 53 summarizes all known regions that have been reported at least once to be non- or only minimally deleterious for the corresponding carriers. More detailed information is presented in Figures 22 and 28.

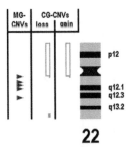

Figure 53 Idiogram of chromosome 22 according to Kosyakova et al. (2009) showing regions that, if deleted or duplicated, have only minor clinical impact. The structure of the figure is explained in the first paragraph of this appendix.

A.23 X-CHROMOSOME

According to GRCh37/hg19 the X-chromosome has a size of 155.3 Mb. Figure 54 summarizes all known regions that have been reported at least once to be non- or only minimally deleterious for the corresponding carriers. More detailed information is presented in Chapter 2, section 2.7.2.2; Chapter 4, section 4.7; and Chapter 5, sections 5.1.2.23, 5.2.1.23, and 5.2.2.23.

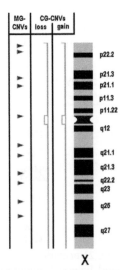

Figure 54 Idiogram of the X chromosome according to Kosyakova et al. (2009) showing regions that, if deleted or duplicated, have only minor clinical impact. The structure of the figure is explained in the first paragraph of this appendix.

A.24 Y-CHROMOSOME

According to GRCh37/hg19 the Y-chromosome has a size of 59.4 Mb. Figure 55 summarizes all known regions that have been reported at least once to be non- or only minimally deleterious for the corresponding carriers. More detailed information is presented in Chapter 2, section 2.7.2.1; Chapter 4, section 4.7; and Chapter 5, sections 5.1.2.24, 5.2.1.24, and 5.2.2.24.

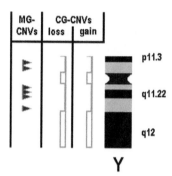

Figure 55 Idiogram of the Y chromosome according to Kosyakova et al. (2009) showing regions that, if deleted or duplicated, have only minor clinical impact. The structure of the figure is explained in the first paragraph of this appendix.

A.25 SHORT ANALYSIS OF THE SUMMARY OF CG-CNVS BY CHROMOSOME

It is obvious and not surprising that all heterochromatic regions of the human genome may be hit by CG-CNVs. For euchromatic CG-CNVs the following correlations may be found when analyzing the figures provided in sections A.1 through A.22; gonosomes are left out here as they may be entirely present or absent without too severe clinical consequences:

- GTG-dark bands spanned by potential deletions: 40/219 (18.3%)
- GTG-light bands spanned by potential deletions: 60/240 (25.0%)
- GTG-dark bands spanned by potential duplications: 74/219 (33.8%)
- GTG-light bands spanned by potential duplications: 97/240 (40.4%)
- GTG-dark bands spanned by potential duplications AND deletions in common: 23/219 (10.5%)
- GTG-light bands spanned by potential duplications AND deletions in common: 29/240 (12.1%)
- Colocalization of potential CG-CNVs with the most frequent MG-CNVs: 80/157 (51%)

According to this data, surprisingly large amounts of the autosomal genome can be subject to CG-CNVs. At an approximately 550 band per haploid karyotype level, approximately 22% of the euchromatic bands potentially may be hit by deletions and about 37% of them by duplications. Such events may not happen at the same time: approximately 11% of the corresponding bands have been observed already as deleted or duplicated in different persons. Overall, this observation fits to the general assessment that euchromatic gain of copy numbers can be more easily tolerated by the genome than loss.

Finally, the colocalization of more than 50% of the most frequently occurring MG-CNVs and the CG-CNVs is striking, even though not that astounding.

1000 Genomes Project Consortium, Abecasis, G.R., Auton, A., Brooks, L.D., DePristo, M.A., Durbin, R.M., Handsaker, R.E., Kang, H.M., Marth, G.T., McVean, G.A., 2012. An integrated map of genetic variation from 1,092 human genomes. Nature 491, 56–65.

Acar, H., Cora, T., Erkul, I., 1999. Coexistence of inverted Y, chromosome 15p+ and abnormal phenotype. Genet. Couns. 10, 163–170.

Acar, H., Yildirim, M.S., Kaynak, M., 2002. Reliability and efficiency of interphase-fish with alpha-satellite probe for detection of aneuploidy. Genet. Couns. 13, 11–17.

Adhvaryu, S.G., Rawal, U.M., 1991. C-band heterochromatin variants in individuals with neoplastic disorders: Carcinoma of breast and ovary. Neoplasma 38, 379–384.

Aguilar, L., Lisker, R., Ruz, L., Mutchinick, O., 1981. Constitutive heterochromatin polymorphisms in patients with malignant diseases. Cancer 47, 2437–2432.

Ait-Allah, A.S., Ming, P.-M.L., Salem, H.T., Reece, E.A., 1997. The clinical importance of pericentric inversion of chromosome 9 in prenatal diagnosis. J. Mat-Fet. Investig. 7, 126–128.

Akkari, Y., Lawce, H., Kelson, S., Smith, C., Davis, C., Boyd, L., Magenis, R.E., Olson, S., 2005. Y chromosome heterochromatin of differing lengths in two cell populations of the same individual. Prenat. Diagn. 25, 304–306.

Albers, C.A., Paul, D.S., Schulze, H., Freson, K., Stephens, J.C., Smethurst, P.A., Jolley, J.D., Cvejic, A., Kostadima, M., Bertone, P., Breuning, M.H., Debili, N., Deloukas, P., Favier, R., Fiedler, J., Hobbs, C.M., Huang, N., Hurles, M.E., Kiddle, G., Krapels, I., Nurden, P., Ruivenkamp, C.A., Sambrook, J.G., Smith, K., Stemple, D.L., Strauss, G., Thys, C., van Geet, C., Newbury-Ecob, R., Ouwehand, W.H., Ghevaert, C., 2012. Compound inheritance of a low-frequency regulatory SNP and a rare null mutation in exon-junction complex subunit RBM8A causes TAR syndrome. Nat. Genet. 44, 435–439, S1–2.

Alexandrov, I.A., Mashkova, T.D., Akopian, T.A., Medvedev, L.I., Kisselev, L.L., Mitkevich, S.P., Yurov, Y.B., 1991. Chromosome-specific alpha satellites: Two distinct families on human chromosome 18. Genomics 11, 15–23.

Alexandrov, I., Kazakov, A., Tumeneva, I., Shepelev, V., Yurov, Y., 2001. Alpha-satellite DNA of primates: Old and new families. Chromosoma 110, 253–266.

Alitalo, T., Tiihonen, J., Hakola, P., de la Chapelle, A., 1988. Molecular characterization of a Y;15 translocation segregating in a family. Hum. Genet. 79, 29–35.

Alkhalaf, M., Verghese, L., Muharib, N., 2002. A cytogenetic study of Kuwaiti couples with infertility and reproductive disorders: Short arm deletion of chromosome 21 is associated with male infertility. Ann. Genet. 45, 147–149.

Allen, T.L., Brothman, A.R., Carey, J.C., Chance, P.F., 1996. Cytogenetic and molecular analysis in trisomy 12p. Am. J. Med. Genet. 63, 250–256.

Aller, V., Abrisqueta, J.A., Pérez-Castillo, A., del Mazo, J., Martín-Lucas, M.A., de Torres, M.L., 1979. Trisomy 10p due to a de novo t(10p;13p). Hum. Genet. 46, 129–134.

Al-Saffar, M., Lemyre, E., Koenekoop, R., Duncan, A.M., Der Kaloustian, V.M., 2000. Phenotype of a patient with pure partial trisomy 2p(p23->pter). Am. J. Med. Genet. 94, 428–432.

Andersson, M., Page, D.C., Pettay, D., Subrt, I., Turleau, C., de Grouchy, J., de la Chapelle, A.Y., 1988. autosome translocations and mosaicism in the aetiology of 45, X maleness: Assignment of fertility factor to distal Yq11. Hum. Genet. 79, 2–7.

Annerén, G., Gustavson, K.H., 1982. A boy with proximal trisomy 15 and a male foetus with distal trisomy 15 due to a familial 13p;15q translocation. Clin. Genet. 22, 16–21.

Arn, P.H., Younie, L., Russo, S., Zackowski, J.L., Mankinen, C., Estabrooks, L., 1995. Reproductive outcome in 3 families with a satellited chromosome 4 with review of the literature. Am. J. Med. Genet. 57, 420–424.

Arnold, J., 1879. Über feinere Strukturen der Zelle unter normalen und pathologischen Bedingungen. Virchows. Arch. Path. Anat. 77, 181–206.

Antonacci, R., Rocchi, M., Archidiacono, N., Baldini, A., 1995. Ordered mapping of three alpha satellite DNA subsets on human chromosome 22. Chromosome Res. 3, 124–127.

Ashton-Prolla, P., Gershin, I.F., Babu, A., Neu, R.L., Zinberg, R.E., Willner, J.P., Desnick, R.J., Cotter, P.D., 1997. Prenatal diagnosis of a familial interchromosomal insertion of Y chromosome heterochromatin. Am. J. Med. Genet. 73, 470–473.

Atkin, N.B., Baker, M.C., 1995. Ectopic nucleolar organizer regions. A common anomaly revealed by Ag-NOR staining of metaphases from nine cancers. Cancer Genet. Cytogenet. 85, 129–132.

Babu, A., Verma, R.S., 1986. Characterization of human chromosomal constitutive heterochromatin. Can. J. Genet. Cytol. 28, 631–644.

Babu, A., Macera, M.J., Verma, R.S., 1986. Intensity heteromorphisms of human chromosome 15p by DA/DAPI technique. Hum. Genet. 73, 298–300.

Babu, A., Agarwal, A.K., Verma, R.S., 1988. A new approach in recognition of hetero- chromatic regions of human chromosomes by means of restriction endonucleases. Am. J. Hum. Genet. 42, 60–65.

Babu, V.R., Roberson, J.R., Van Dyke, D.L., Weiss, L., 1987. Interstitial deletion of 4q35 in a familial satellited 4q in a child with developmental delay. Am. J. Hum. Genet. 41 (Suppl.), 113.

Bailey, J.A., Yavor, A.M., Massa, H.F., Trask, B.J., Eichler, E.E., 2001. Segmental dupli- cations: Organization and impact within the current human genome project assembly. Genome Res. 11, 1005–1017.

Bajnóczky, K., Meggyessy, V., 1985. Coincidence of paternal 13pYq translocation and maternal increased 13p NOR activity in a child with arthrogryposis and other mal- formations. Acta. Paediatr. Hung. 26, 151–156.

Bakhoum, S.F., Compton, D.A., 2012. Chromosomal instability and cancer: A complex relationship with therapeutic potential. J. Clin. Invest. 122, 1138–1143.

Balciuniene, J., Feng, N., Iyadurai, K., Hirsch, B., Charnas, L., Bill, B.R., Easterday, M.C., Staaf, J., Oseth, L., Czapansky-Beilman, D., Avramopoulos, D., Thomas, G.H., Borg, A., Valle, D., Schimmenti, L.A., Selleck, S.B., 2007. Recurrent 10q22-q23 deletions: A genomic disorder on 10q associated with cognitive and behavioral abnormalities. Am. J. Hum. Genet. 80, 938–947.

Baldwin, E.L., May, L.F., Justice, A.N., Martin, C.L., Ledbetter, D.H., 2008. Mechanisms and consequences of small supernumerary marker chromosomes: From Barbara McClintock to modern genetic-counseling issues. Am. J. Hum. Genet. 82, 398–410.

Balkan, W., Burns, K., Martin, R.H., 1983. Sperm chromosome analysis of a man het- erozygous for a pericentric inversion of chromosome 3. Cytogenet. Cell Genet. 35, 295–297.

Ballif, B.C., Kashork, C.D., Shaffer, L.G., 2000. The promise and pitfalls of telomere region-specific probes. Am. J. Hum. Genet. 67, 1356–1359.

Baker, E., Hinton, L., Callen, D.F., Haan, E.A., Dobbie, A., Sutherland, G.R., 2002. A familial cryptic subtelomeric deletion 12p with variable phenotypic effect. Clin. Genet. 61, 198–201.

Bandyopadhyay, R., McQuillan, C., Page, S.L., Choo, K.H., Shaffer, L.G., 2001. Identification and characterization of satellite III subfamilies to the acrocentric chromosomes. Chromosome Res. 9, 223–233.

Barber, J.C., 2005. Directly transmitted unbalanced chromosome abnormalities and euchromatic variants. J. Med. Genet. 42, 609–629.

Barber, J.C., 2013. http://www.ngrl.org.uk/wessex/collection/.

Barber, J.C., Reed, C.J., Dahoun, S.P., Joyce, C.A., 1999. Amplification of a pseudogene cassette underlies euchromatic variation of 16p at the cytogenetic level. Hum. Genet. 104, 211–218.

Barber, J.C., Maloney, V., Hollox, E.J., Stuke-Sontheimer, A., du Bois, G., Daumiller, E., Klein-Vogler, U., Dufke, A., Armour, J.A., Liehr, T., 2005. Duplications and copy number variants of 8p23.1 are cytogenetically indistinguishable but distinct at the molecular level. Eur. J. Hum. Genet. 13, 1131–1136.

Barber, J.C., Zhang, S., Friend, N., Collins, A.L., Maloney, V.K., Hastings, R., Farren, B., Barnicoat, A., Polityko, A.D., Rumyantseva, N.V., Starke, H., Ye, S., 2006. Duplications of proximal 16q flanked by heterochromatin are not euchromatic variants and show no evidence of heterochromatic position effect. Cytogenet. Genome Res. 114, 351–358.

Barber, J.C., Maloney, V.K., Bewes, B., Wakeling, E., 2006. Deletions of 2q14 that include the homeobox engrailed 1 (EN1) transcription factor are compatible with a normal phenotype. Eur. J. Hum. Genet. 14, 739–743.

Barber, J.C., Maloney, V.K., Kirchhoff, M., Thomas, N.S., Boyle, T.A., Castle, B., 2007. Transmitted duplication of 12q21.32-12q22 includes 48 genes and has no apparent phenotypic consequences. Am. J. Med. Genet. A 143, 615–618.

Barber, J.C., Brasch-Andersen, C., Maloney, V.K., Huang, S., Bateman, M.S., Graakjaer, J., Heinl, U.D., Fagerberg, C., 2013. A novel pseudo-dicentric variant of 16p11.2-q11.2 contains euchromatin from 16p11.2-p11.1 and resembles pathogenic duplications of proximal 16q. Cytogenet. Genome Res. 139, 59–64.

Barber, J.C., Hall, V., Maloney, V.K., Huang, S., Roberts, A.M., Brady, A.F., Foulds, N., Bewes, B., Volleth, M., Liehr, T., Mehnert, K., Bateman, M., White, H., 2013. 16p11.2-p12.2 duplication syndrome; a genomic condition differentiated from euchromatic variation of 16p11.2. Eur. J. Hum. Genet. 21, 182–189.

Bardhan, S., Singh, D.N., Davis, K., 1981. Polymorphism in chromosome 4. Clin. Genet. 20, 44–47.

Bartsch, O., Kalbe, U., Ngo, T.K., Lettau, R., Schwinger, E., 1990. Clinical diagnosis of partial duplication 7q. Am. J. Med. Genet. 37, 254–257.

Bartsch, O., König, U., Petersen, M.B., Poulsen, H., Mikkelsen, M., Palau, F., Prieto, F., Schwinger, E., 1993. Cytogenetic, FISH and DNA studies in 11 individuals from a family with two siblings with dup(21q) Down syndrome. Hum. Genet. 92, 127–132.

Bassi, C., Magnani, I., Sacchi, N., Saccone, S., Ventura, A., Rocchi, M., Marozzi, A., Ginelli, E., Meneveri, R., 2000. Molecular structure and evolution of DNA sequences located at the alpha satellite boundary of chromosome 20. Gene 256, 43–50.

Bauld, R., Ellis, P.M., 1984. A satellited chromosome 2. J. Med. Genet. 21, 54.

Bauchinger, M., Schmid, E., 1970. A case with balanced (14p+; 15p minus)-translocation. Humangenetik 8, 312–320.

Bayless-Underwood, L., Cho, S., Ward, B., Robinson, A., 1983. Two cases of prenatal diagnosis of a satellited Yq chromosome. Clin. Genet. 24, 359–364.

Bedoyan, J.K., Flore, L.A., Alkatib, A., Ebrahim, S.A., Bawle, E.V., 2004. Transmission of ring chromosome 13 from a mother to daughter with both having a 46,XX, r(13)(p13q34) karyotype. Am. J. Med. Genet. A 129A, 316–320.

Begovic, D., Hitrec, V., Lasan, R., Letica, L., Baric, I., Sarnavka, V., Galic, S., 1998. Partial trisomy 13 in an infant with a mild phenotype: Application of fluorescence in situ hybridization in cytogenetic syndromes. Croat. Med. J. 39, 212–215.

Bendavid, C., Pasquier, L., Watrin, T., Morcel, K., Lucas, J., Gicquel, I., Dubourg, C., Henry, C., David, V., Odent, S., Levêque, J., Pellerin, I., Guerrier, D., 2007. Phenotypic variability of a 4q34->qter inherited deletion: MRKH syndrome in the daughter, cardiac defect and Fallopian tube cancer in the mother. Eur. J. Med. Genet. 50, 66–72.

Benítez, J., Ramos, C., García Quesada, L., Alio, J., 1979. Corneal distrophy of Groenouw type I and chromosomic delection (22p-) in one family. An. Esp. Pediatr. 12, 807–810.

Benzacken, B., Monier-Gavelle, F., Siffroi, J.P., Agbo, P., Chalvon, A., Wolf, J.P., 2001. Acrocentric chromosome polymorphisms: Beware of cryptic translocations. Prenat. Diagn. 21, 96–98.

Berg, J.S., Potocki, L., Bacino, C.A., 2010. Common recurrent microduplication syndromes: Diagnosis and management in clinical practice. Am. J. Med. Genet. A 152A, 1066–1078.

Bernstein, R., Dawson, B., Griffiths, J., 1981. Human inherited marker chromosome 22 short-arm enlargement: Investigation of rDNA gene multiplicity, Ag-band size, and acrocentric association. Hum. Genet. 58, 135–139.

Bertini, V., Valetto, A., Uccelli, A., Tarantino, E., Simi, P., 2004. Ring chromosome 21 and reproductive pattern: A familial case and review of the literature. Fertil. Steril. 2008, 90, e1-5.

Betz, J.L., Behairy, A.S., Rabionet, P., Tirtorahardjo, B., Moore, M.W., Cotter, P.D., 2005. Acquired inv(9): What is its significance? Cancer Genet. Cytogenet. 160, 76–78.

Beverstock, G.C., Klumper, F., Helderman, V.D., Enden, A.T., 1997. Yet another variation on the theme of chromosome 18 heteromorphisms? Prenat. Diagn. 17, 585–586.

Bisgaard, A.M., Kirchhoff, M., Nielsen, J.E., Brandt, C., Hove, H., Jepsen, B., Jensen, T., Ullmann, R., Skovby, F., 2007. Transmitted cytogenetic abnormalities in patients with mental retardation: Pathogenic or normal variants? Eur. J. Med. Genet. 50, 243–255.

Blancato, J.K., 1996. Re: Cross-hybridization of the chromosome 13/21 alpha satellite DNA probe to chromosome 22 in the prenatal screening of common aneuploidies by FISH. Prenat. Diagn. 16, 769–770.

Bloom, S.E., Goodpasture, C., 1975. An improved technique for selective silver staining of nucleolar organizer regions in human chromosomes. Hum. Genet. 34, 199–206.

Blumberg, B.D., Shulkin, J.D., Rotter, J.I., Mohandas, T., Kaback, M.M., 1982. Minor chromosomal variants and major chromosomal anomalies in couples with recurrent abortion. Am. J. Hum. Genet. 34, 948–960.

Bochkov, N.P., Kuleshov, N.P., Chebotarev, A.N., Alekhin, V.I., Midian, S.A., 1974. Population cytogenetic investigation of newborns in Moscow. Humangenetik 22, 139–152.

Bonfatti, A., Giunta, C., Sensi, A., Gruppioni, R., Rubini, M., Fontana, F., 1993. Heteromorphism of human chromosome 18 detected by fluorescent in situ hybridization. Eur. J. Histochem. 37, 149–154.

Borgaonkar, D.S., Bias, W.B., Chase, G.A., Sadasivan, G., Herr, H.M., Golomb, H.M., Bahr, G.F., Kunkel, L.M., 1973. Identification of a C6-G21 translocation chromosome by the Q-M and Giemsa banding techniques in a patient with Down's syndrome, with possible assignment of Gm locus. Clin. Genet. 4, 53–57.

Bossuyt, P.J., Van Tienen, M.N., De Gruyter, L., Smets, V., Dumon, J., Wauters, J.G., 1995. Incidence of low-fluorescence alpha satellite region on chromosome 21 escaping detection of aneuploidy at interphase by FISH. Cytogenet. Cell Genet. 68, 203–206.

Bouhjar, I.B., Hannachi, H., Zerelli, S.M., Labalme, A., Gmidène, A., Soyah, N., Missaoui, S., Sanlaville, D., Elghezal, H., Saad, A., 2011. Array-CGH study of partial trisomy 9p without mental retardation. Am. J. Med. Genet. A 155A, 1735–1739.

Bourthoumieu, S., Esclaire, F., Terro, F., Brosset, P., Fiorenza, M., Aubard, V., Beguet, M., Yardin, C., 2010. Familial 18 centromere variant resulting in difficulties in interpreting prenatal interphase FISH. Morphologie 94, 68–72.

Boyd, L.J., Livingston, J.S., Brown, M.G., Lawce, H.J., Gilhooly, J.T., Wildin, R.S., Linck, L.M., Magenis, R.E., Pillers, D.A., 2005. Meiotic exchange event within the stalk region of an inverted chromosome 22 results in a recombinant chromosome with duplication of the distal long arm. Am. J. Med. Genet. A 138, 355–360.

Brito-Babapulle, V., 1981. Lateral asymmetry in human chromosomes 1, 3, 4, 15, and 16. Cytogenet. Cell Genet. 29, 198–202.

Brothman, A.R., Schneider, N.R., Saikevych, I., Cooley, L.D., Butler, M.G., Patil, S., Mascarello, J.T., Rao, K.W., Dewald, G.W., Park, J.P., Persons, D.L., Wolff, D.J., Vance, G.H., 2006. Cytogenetics Resource Committee, College of American Pathologists/American College of Medical Genetics. Cytogenetic heteromorphisms: survey results and reporting practices of giemsa-band regions that we have pondered for years. Arch. Pathol. Lab. Med. 130, 947–949.

Brown, W.R., MacKinnon, P.J., Villasanté, A., Spurr, N., Buckle, V.J., Dobson, M.J., 1990. Structure and polymorphism of human telomere-associated DNA. Cell 63, 119–132.

Bruder, C.E., Piotrowski, A., Gijsbers, A.A., Andersson, R., Erickson, S., Diaz de Ståhl, T., Menzel, U., Sandgren, J., von Tell, D., Poplawski, A., Crowley, M., Crasto, C., Partridge, E.C., Tiwari, H., Allison, D.B., Komorowski, J., van Ommen, G.J., Boomsma, D.I., Pedersen, N.L., den Dunnen, J.T., Wirdefeldt, K., Dumanski, J.P., 2008. Phenotypically concordant and discordant monozygotic twins display different DNA copy-number-variation profiles. Am. J. Hum. Genet. 82, 763–771.

Bucksch, M., Ziegler, M., Kosayakova, N., Mulatinho, M.V., Llerena Jr., J.C., Morlot, S., Fischer, W., Polityko, A.D., Kulpanovich, A.I., Petersen, M.B., Belitz, B., Trifonov, V., Weise, A., Liehr, T., Hamid, A.B., 2012. A new multicolor fluorescence in situ hybridization probe set directed against human heterochromatin: HCM-FISH. J. Histochem. Cytochem. 60, 530–536.

Buckton, K.E., O'Riordan, M.L., Ratcliffe, S., Slight, J., Mitchell, M., McBeath, S., Keay, A.J., Barr, D., Short, M., 1980. A G-band study of chromosomes in liveborn infants. Ann. Hum. Genet. 43, 227–239.

Buiting, K., Dittrich, B., Dworniczak, B., Lerer, I., Abeliovich, D., Cottrell, S., Temple, I.K., Harvey, J.F., Lich, C., Gross, S., Horsthemke, B., 1999. A 28-kb deletion spanning D15S63 (PW71) in five families: A rare neutral variant? Am. J. Hum. Genet. 65, 1588–1594.

Buño, I., Fernández, J.L., López-Fernández, C., Díez-Martín, J.L., Gosálvez, J., 2001. Sau3A in situ digestion of human chromosome 3 pericentrometric heterochromatin. I. Differential digestion of alpha-satellite and satellite 1 DNA sequences. Genome 44, 120–127.

Burk, R.D., Stamberg, J., Young, K.E., Smith, K.D., 1983. Use of repetitive DNA for diagnosis of chromosomal rearrangements. Hum. Genet. 64, 339–342.

Butomo, I.V., Prozorova, M.V., Khitrikova, L.E., 1984. Multiple chromosome aberrations in 3 generations of a family and Down's syndrome resulting from partial trisomy of chromosome 21 (q21–q22). Tsitol. Genet. 18, 223–228.

Buys, C.H., Anders, G.J., Borkent-Ypma, J.M., Blenkers-Platter, J.A., van der Hoek-van der Veen, A.Y., 1979. Familial transmission of a translocation Y/14. Hum. Genet. 53, 125–127.

Caglayan, A.O., Ozgun, M.T., Demiryilmaz, F., Ozyazgan, I., 2009. Can heterochromatin polymorphism of chromosome 6 affect fertility? Genet. Couns. 20, 203–206.

Caglayan, A.O., Gumus, H., 2010. Autism with del15p.11.1: Case report with a new cytogenetic finding. Genet. Couns. 21, 199–204.

Caliebe, A., Waltz, S., Jenderny, J., 1997. Mild phenotypic manifestations of terminal deletion of the long arm of chromosome 4: Clinical description of a new patient. Clin. Genet. 52, 116–119.

Callen, D.F., Eyre, H.J., Ringenbergs, M.L., 1990. A dicentric variant of chromosome 6: Characterization by use of in situ hybridisation with the biotinylated probe p308. Clin. Genet. 37, 81–83.

Callen, D.F., Eyre, H., Lane, S., Shen, Y., Hansmann, I., Spinner, N., Zackai, E., McDonald-McGinn, D., Schuffenhauer, S., Wauters, J., Van Thienen, M.N., Van Roy, B., Sutherland, G.R., Haan, E.A., 1993. High resolution mapping of interstitial long arm deletions of chromosome 16: Relationship to phenotype. J. Med. Genet. 30, 828–832.

Canales, C.P., Walz, K., 2011. Copy number variation and susceptibility to complex traits. EMBO Mol. Med. 3, 1–4.

Chantot-Bastaraud, S., Siffroi, J.P., Berkane, N., Heim, N., Herve, F., Uzan, S., Vendrely, E., 2003. Prenatal diagnosis of a large centromeric heteromorphism of chromosome 12: Implications for genetic counseling. Fetal. Diagn. Ther. 18, 111–113.

Carine, K., Jacquemin-Sablon, A., Waltzer, E., Mascarello, J., Scheffler, I.E., 1989. Molecular characterization of human minichromosomes with centromere from chromosome 1 in human-hamster hybrid cells. Somat. Cell Mol. Genet. 15, 445–460.

Charlieu, J.P., Laurent, A.M., Orti, R., Viegas-Péquignot, E., Bellis, M., Roizès, G., 1993. A 37-kb fragment common to the pericentromeric region of human chromosomes 13 and 21 and to the ancestral inactive centromere of chromosome 2. Genomics 15, 576–581.

Chia, N.L., Bousfield, L.R., Johnson, B.H., 1987. A case report of a de novo tandem duplication (5p)(p14—pter). Clin. Genet. 31, 65–69.

Carmany, E.P., Bawle, E.V., 2011. Microduplication of 4p16.3 due to an unbalanced translocation resulting in a mild phenotype. Am. J. Med. Genet. A 155A, 819–824.

Carnevale, A., Ibañez, B.B., del Castillo, V., 1976. The segregation of C-band polymorphisms on chromosomes 1, 9, and 16. Am. J. Hum. Genet. 28, 412–416.

Caspersson, T., Farber, S., Foley, G.E., Kudynowski, J., Modest, E.J., Simonsson, E., Wagh, U., Zech, L., 1968. Chemical differentiation along metaphase chromosomes. Exp. Cell Res. 49, 219–222.

Cavalli, I.J., Mattevi, M.S., Erdtmann, B., Sbalqueiro, I.J., Maia, N.A., 1985. Equivalence of the total constitutive heterochromatin content by an interchromosomal compensation in the C band sizes of chromosomes 1, 9, 16, and Y in Caucasian and Japanese individuals. Hum. Hered. 35, 379–387.

Ceccarini, C., Sinibaldi, L., Bernardini, L., De Simone, R., Mingarelli, R., Novelli, A., Dallapiccola, B., 2007. Duplication 18q21.31-q22.2. Am. J. Med. Genet. A 143, 343–348.

Chatzimeletiou, K., Taylor, J., Marks, K., Grudzinskas, J.G., Handyside, A.H., 2006. Paternal inheritance of a 16qh-polymorphism in a patient with repeated IVF failure. Reprod. Biomed. Online 13, 864–867.

Chen, L.L., Carmichael, G.G., 2010. Long noncoding RNAs in mammalian cells: What, where, and why? Wiley Interdiscip. Rev. RNA 1, 2–21.

Chen, T.R., Kao, M.L., Marks, J., Chen, Y.Y., 1981. Polymorphic variants in human chromosome 15. Am. J. Med. Genet. 9, 61–66.

Chen, C.P., Devriendt, K., Chern, S.R., Lee, C.C., Wang, W., Lin, S.P., 2000. Prenatal diagnosis of inherited satellited non-acrocentric chromosomes. Prenat. Diagn. 20, 384–389.

Chen, C.P., Lin, S.P., 2003. Distal 10q trisomy associated with bilateral hydronephrosis in infancy. Genet. Couns. 14, 359–362.

Chen, C.P., Chern, S.R., Lee, C.C., Chen, W.L., Wang, W., 2004. Prenatal diagnosis of interstitially satellited 6p. Prenat. Diagn. 24, 430–433.

Chen, Y., Chen, G., Lian, Y., Gao, X., Huang, J., Qiao, J., 2007. A normal birth following preimplantation genetic diagnosis by FISH determination in the carriers of der(15) t(Y;15)(Yq12;15p11) translocations: Two case reports. J. Assist. Reprod. Genet. 24, 483–488.

Cheng, Z.Y., Gao, C.S., Xin, X., Fu, S.M., Zhong, W.L., 1989. Molecular cytogenetic study on the case with 14p+ marker chromosome. Yi Chuan Xue Bao 16, 331–334.

Chen-Shtoyerman, R., Josefsberg Ben-Yehoshua, S., Nissani, R., Rosensaft, J., Appelman, Z., 2012. A prevalent Y;15 translocation in the Ethiopian Beta Israel community in Israel. Cytogenet. Genome Res. 136, 171–174.

Chia, N.L., Bousfield, L.R., Poon, C.C., Trudinger, B.J., 1988. Trisomy (1q)(q42—qter): confirmation of a syndrome. Clin. Genet. 34, 224–229.

Choo, K.H., Brown, R., Webb, G., Craig, I.W., Filby, R.G., 1987. Genomic organization of human centromeric alpha satellite DNA: Characterization of a chromosome 17 alpha satellite sequence. DNA 6, 297–305.

Choo, K.H., Earle, E., Vissel, B., Filby, R.G., 1990. Identification of two distinct sub-families of alpha satellite DNA that are highly specific for human chromosome 15. Genomics 7, 143–151.

Choo, K.H., Vissel, B., Nagy, A., Earle, E., Kalitsis, P., 1991. A survey of the genomic distribution of alpha satellite DNA on all the human chromosomes, and derivation of a new consensus sequence. Nucleic Acids Res. 19, 1179–1182.

Choo, K.H., Earle, E., Vissel, B., Kalitsis, P., 1992. A chromosome 14-specific human satellite III DNA subfamily that shows variable presence on different chromosomes 14. Am. J. Hum. Genet. 50, 706–716.

Claussen, U., Michel, S., Mühlig, P., Westermann, M., Grummt, U.W., Kromeyer-Hauschild, K., Liehr, T., 2002. Demystifying chromosome preparation and the implications for the concept of chromosome condensation during mitosis. Cytogenet. Genome Res. 98, 136–146.

Cockwell, A.E., Jacobs, P.A., Beal, S.J., Crolla, J.A., 2003. A study of cryptic terminal chromosome rearrangements in recurrent miscarriage couples detects unsuspected acrocentric pericentromeric abnormalities. Hum. Genet. 112, 298–302.

Cockwell, A.E., Jacobs, P.A., Crolla, J.A., 2007. Distribution of the D15Z1 copy number polymorphism. Eur. J. Hum. Genet. 15, 441–445.

Codina-Pascual, M., Navarro, J., Oliver-Bonet, M., Kraus, J., Speicher, M.R., Arango, O., Egozcue, J., Benet, J., 2006. Behaviour of human heterochromatic regions during the synapsis of homologous chromosomes. Hum. Reprod. 21, 1490–1497.

Collin, A., Sladkevicius, P., Soller, M., 2009. False-positive prenatal diagnosis of trisomy 18 by interphase FISH: Hybridization of chromosome 18 alpha-satellite probe (D18Z1) to chromosome 2. Prenat. Diagn. 29, 1279–1281.

Colls, P., Sandalinas, M., Pagidas, K., Munné, S., 2004. PGD analysis for aneuploidy in a patient heterozygous for a polymorphism of chromosome 16 (16qh-). Prenat. Diagn. 24, 741–744.

Colnaghi, R., Carpenter, G., Volker, M., O'Driscoll, M., 2011. The consequences of structural genomic alterations in humans: Genomic disorders, genomic instability and cancer. Semin. Cell Dev. Biol. 22, 875–885.

Conrad, D.F., Pinto, D., Redon, R., Feuk, L., Gokcumen, O., Zhang, Y., Aerts, J., Andrews, T.D., Barnes, C., Campbell, P., Fitzgerald, T., Hu, M., Ihm, C.H., Kristiansson, K., Macarthur, D.G., Macdonald, J.R., Onyiah, I., Pang, A.W.,

Robson, S., Stirrups, K., Valsesia, A., Walter, K., Wei, J., Wellcome Trust Case Control Consortium, Tyler-Smith, C., Carter, N.P., Lee, C., Scherer, S.W., Hurles, M.E., 2010. Origins and functional impact of copy number variation in the human genome. Nature 464, 704–712.

Cooper, K.F., Fisher, R.B., Tyler-Smith, C., 1993. The major centromeric array of alphoid satellite DNA on the human Y chromosome is non-palindromic. Hum. Mol. Genet. 2, 1267–1270.

Cosper, R., Hicks, L.C., Finley, S.C., Davis, R.O., Carroll, A.J., 1985. Familial insertion of nucleolar organizer regions and centromere material into the long arm of 11. Am. J. Hum. Genet. 37, A89.

Conte, R.A., Luke, S., Verma, R.S., 1992. Molecular characterization of "inverted" pericentromeric heterochromatin of chromosome 3. Histochemistry 97, 509–510.

Conte, R.A., Mathews, T., Kleyman, S.M., Verma, R.S., 1996. Molecular characterization of 21p- variant chromosome. Clin. Genet. 50, 103–105.

Conte, R.A., Kleyman, S.M., Laundon, C., Verma, R.S., 1997. Characterization of two extreme variants involving the short arm of chromosome 22: Are they identical? Ann. Genet. 40, 145–149.

Couturier-Turpin, M.H., Ingster, O., Salat-Baroux, J., Feldmann, G., 1994. Report of a family case of satellited Y chromosome associated with a severe oligoasthenoteratospermia. A review of the literature. Ann. Genet. 37, 200–206.

Craig-Holmes, A.P., Moore, F.B., Shaw, M.W., 1973. Polymorphism of human C-band heterochromatin. I. Frequency of variants. Am. J. Hum. Genet. 25, 181–192.

Craig-Holmes, A.P., Moore, F.B., Shaw, M.W., 1975. Polymporphism of human C-band heterochromatin. II. Family studies with suggestive evidence for somatic crossing over. Am. J. Hum. Genet. 27, 178–189.

Crossen, P.E., 1975. Variation in the centromeric banding of chromosome 19. Clin. Genet. 8, 218–222.

Cuoco, C., Ronchetto, P., Gimelli, S., Béna, F., Divizia, M.T., Lerone, M., Mirabelli-Badenier, M., Mascaretti, M., Gimelli, G., 2011. Microarray based analysis of an inherited terminal 3p26.3 deletion, containing only the CHL1 gene, from a normal father to his two affected children. Orphanet. J. Rare Dis. 6, 12.

D'Aiuto, L., Antonacci, R., Marzella, R., Archidiacono, N., Rocchi, M., 1993. Cloning and comparative mapping of a human chromosome 4-specific alpha satellite DNA sequence. Genomics 18, 230–235.

Dallapiccola, B., Brinchi, V., Magnani, M., Dacha, M., 1980. Identification of the origin of a 22p+ chromosome by triplex dosage effect of LDH B, GAPHD, TPI and ENO2. Ann. Genet. 23, 111–113.

Daniel, A., Darmanian, A., Peters, G., Goodwin, L., Hort, J.R., 2007. An innocuous duplication of 11.2 Mb at 13q21 is gene poor: Sub-bands of gene paucity and pervasive CNV characterize the chromosome anomalies. Am. J. Med. Genet. A 143A, 2452–2459.

De Cabo, S.F., Ludeña, P., Velázquez, M., Sentis, C., Fernández-Piqueras, J., 1991. Cryptic variants of acrocentric human chromosomes as analysed by restriction endonucleases. Genetica 83, 203–206.

de Carvalho, A.F., da Silva Bellucco, F.T., Kulikowski, L.D., Toralles, M.B., Melaragno, M.I., 2008. Partial 5p monosomy or trisomy in 11 patients from a family with a t(5;15)(p13.3;p12) translocation. Hum. Gene. 124, 387–392.

de Chieri, P., Spatuzza, E., Bonich, J.M., 1978. Brother and sister with trisomy 10p. 46, XY,(22p+)mat; 46,XX,(22p+)mat. Hum. Genet. 45, 71–75.

Dekaban, A.S., Bender, M.A., Economos, G.E., 1963. Chromosome studies in mongoloids and their families. Cytogenetics 2, 61–75.

de la Chapelle, A., Schröder, J., Stenstrand, K., Fellman, J., Herva, R., Saarni, M., Anttolainen, I., Tallila, I., Tervilä, L., Husa, L., Tallqvist, G., Robson, E.B., Cook, P.J., Sanger, R., 1974. Pericentric inversions of human chromosomes 9 and 10. Am. J. Hum. Genet. 26, 746–766.

De la Fuente-Cortés, B.E., Cerda-Flores, R.M., Dávila-Rodríguez, M.I., García-Vielma, C., De la Rosa Alvarado, R.M., Cortés-Gutiérrez, E.I., 2009. Chromosomal abnormalities and polymorphic variants in couples with repeated miscarriage in Mexico. Reprod. Biomed. Online 18, 543–548.

de la Puente, A., Velasco, E., Pérez Jurado, L.A., Hernández-Chico, C., van de Rijke, F.M., Scherer, S.W., Raap, A.K., Cruces, J., 1998. Analysis of the monomeric alphoid sequences in the pericentromeric region of human chromosome 7. Cytogenet. Cell. Genet. 83, 176–181.

De los Cobos, L., Ligia, A.P., 1981. Preferential segregation of chromosome 21p- in 3 generations. Ultimate role in non-disjunction (apropos of a case of trisomy 21 in this family). J. Genet. Hum. 28, 201–206.

De Marchi, M., Zuffardi, O., Carozzi, F., Schmid, W., Carbonara, A.O., 1977. Partial trisomy 3q in a newborn female. Ric. Clin. Lab. 7, 225–232.

De Pater, J.M., Van Tintelen, J.P., Stigter, R., Brouwers, H.A., Scheres, J.M., 2000. Precarious acrocentric short arm in prenatal diagnosis: No chromosome 14 polymorphism, but trisomy 17p. Genet. Couns. 11, 241–247.

de Pater, J.M., Brocker-Vriends, A.H., Verschuren, M., Linders, D.A., Hansson, K.B., 2006. Misleading variant chromosome 12 in prenatal diagnosis. Prenat. Diagn. 26, 587–588.

de Ravel, T., Aerssens, P., Vermeesch, J.R., Fryns, J.P., 2005. Trisomy of chromosome 16p13.3 due to an unbalanced insertional translocation into chromosome 22p13. Eur. J. Med. Genet. 48, 355–359.

Dev, V.G., Byrne, J., Bunch, G., 1979. Partial translocation of NOR and its activity in a balanced carrier and in her cri-du-chat fetus. Hum. Genet. 51, 277–280.

Devilee, P., Slagboom, P., Cornelisse, C.J., Pearson, P.L., 1986. Sequence heterogeneity within the human alphoid repetitive DNA family. Nucleic Acids Res. 14, 2059–2073.

Devilee, P., Kievits, T., Waye, J.S., Pearson, P.L., Willard, H.F., 1988. Chromosome-specific alpha satellite DNA: Isolation and mapping of a polymorphic alphoid repeat from human chromosome 10. Genomics 3, 1–7.

Di Bella, M.A., Calì, F., Seidita, G., Mirisola, M., Ragusa, A., Ragalmuto, A., Galesi, O., Elia, M., Greco, D., Zingale, M., Gambino, G., D'Anna, R.P., Regan, R., Carbone, M.C., Gallo, A., Romano, V., 2006. Screening of subtelomeric rearrangements in autistic disorder: Identification of a partial trisomy of 13q34 in a patient bearing a 13q;21p translocation. Am. J. Med. Genet. B Neuropsychiatr. Genet. 141B, 584–590.

Djalali, M., Steinbach, P., Bullerdiek, J., Holmes-Siedle, M., Verschraegen-Spae, M.R., Smith, A., 1986. The significance of pericentric inversions of chromosome 2. Hum. Genet. 72, 32–36.

Docherty, Z., Bowser-Riley, S.M., 1984. A rare heterochromatic variant of chromosome 4. J. Med. Genet. 21, 470–472.

Doneda, L., Magnani, I., Tibiletti, M.G., Dalprà, L., Larizza, L., 1992. Different phenotypes in two cases of an apparently identical familial (Yq;13p) translocation. Hum. Reprod. 7, 495–499.

Doneda, L., Gandolfi, P., Nocera, G., Larizza, L., 1998. A rare chromosome 5 heterochromatic variant derived from insertion of 9qh satellite 3 sequences. Chromosome Res. 6, 411–414.

Doyle, C.T., 1976. The cytogenetics of 90 patients with idiopathic mental retardation/malformation syndromes and of 90 normal subjects. Hum. Genet. 33, 131–146.

Dufke, A., Singer, S., Borell-Kost, S., Stotter, M., Pflumm, D.A., Mau-Holzmann, U.A., Starke, H., Mrasek, K., Enders, H., 2006. De novo structural chromosomal imbalances: Molecular cytogenetic characterization of partial trisomies. Cytogenet. Genome Res. 114, 342–250.

Durkin, K., Coppieters, W., Drögemüller, C., Ahariz, N., Cambisano, N., Druet, T., Fasquelle, C., Haile, A., Horin, P., Huang, L., Kamatani, Y., Karim, L., Lathrop, M., Moser, S., Oldenbroek, K., Rieder, S., Sartelet, A., Sölkner, J., Stålhammar, H., Zelenika, D., Zhang, Z., Leeb, T., Georges, M., Charlier, C., 2012. Serial translocation by means of circular intermediates underlies colour sidedness in cattle. Nature 482, 81–84.

Duval, A., Feneux, D., Sutton, L., Tchernia, G., Léonard, C., 2000. Spurious monosomy 7 in leukemia due to centromeric heteromorphism. Cancer Genet. Cytogenet. 119, 67–69.

Earle, E., Dale, S., Choo, K.H., 1989. Amplification of satellite III DNA in an unusually large chromosome 14p+ variant. Hum. Genet. 82, 187–190.

Earle, E., Voullaire, L.E., Hills, L., Slater, H., Choo, K.H., 1992. Absence of satellite III DNA in the centromere and the proximal long-arm region of human chromosome 14: Analysis of a 14p- variant. Cytogenet. Cell Genet. 61, 78–80.

Estabrooks, L.L., Lamb, A.N., Kirkman, H.N., Callanan, N.P., Rao, K.W., 1992. A molecular deletion of distal chromosome 4p in two families with a satellited chromosome 4 lacking the Wolf-Hirschhorn syndrome phenotype. Am. J. Hum. Genet. 51, 971–978.

Eichler, E.E., Budarf, M.L., Rocchi, M., Deaven, L.L., Doggett, N.A., Baldini, A., Nelson, D.L., Mohrenweiser, H.W., 1997. Interchromosomal duplications of the adrenoleukodystrophy locus: A phenomenon of pericentromeric plasticity. Hum. Mol. Genet. 6, 991–1002.

Eisenbarth, I., König-Greger, D., Wöhr, G., Kehrer-Sawatzki, H., Assum, G., 1999. Characterization of an alphoid subfamily located near p-arm sequences on human chromosome 22. Chromosome Res. 7, 65–69.

el-Badramany MH, Farag, T.I., al-Awadi, S.A., Hammad, I.M., Abdelkader, A., Murthy, D.S., 1989. Familial manic-depressive illness with deleted short arm of chromosome 21: Coincidental or causal? Br. J. Psychiatry. 155, 856–857.

Ellis, P.M., West, J.D., West, K.M., Murray, R.S., Coyle, M.C., 1990. Relevance to prenatal diagnosis of the identification of a human Y/autosome translocation by Y-chromosome-specific in situ hybridisation. Mol. Reprod. Dev. 25, 37–41.

Elliot, J., Barnes, I.C.S., 1992. A satellited chromosome 2 detected at prenatal diagnosis. J. Med. Genet. 29, 213.

Emerit, I., de Grouchy, J., German, J., 1968. A short arm deletion of chromosome 13. Ann. Genet. 11, 184–186.

Emerit, I., Noel, B., Thiriet, M., Loubon, M., Quack, B., 1972. Short arm deletion of chromosome 14. Humangenetik 15, 33–38.

Enukashvily, N.I., Donev, R., Waisertreiger, I.S., Podgornaya, O.I., 2007. Human chromosome 1 satellite 3 DNA is decondensed, demethylated and transcribed in senescent cells and in A431 epithelial carcinoma cells. Cytogenet. Genome Res. 118, 42–54.

Eymery, A., Souchier, C., Vourc'h, C., Jolly, C., 2010. Heat shock factor 1 binds to and transcribes satellite II and III sequences at several pericentromeric regions in heat-shocked cells. Exp. Cell Res. 316, 1845–1855.

Engelen, J.J., Marcelis, C.L., Alofs, M.G., Loneus, W.H., Pulles-Heintzberger, C.F., Hamers, A.J., 2001. De novo "pure" partial trisomy (6)(p22.1–>pter) in a chromosome 15 with an enlarged satellite, identified by microdissection. Am. J. Med. Genet. 99, 48–53.

Faivre, L., Morichon-Delvallez, N., Viot, G., Larget-Piet, A., Narcy, F., Turleau, C., Pinson, M.P., Dumez, Y., Munnich, A., Vekemans, M., 1999. Prenatal diagnosis of a satellited non-acrocentric chromosome derived from a maternal translocation (10;13)(p13;p12) and review of literature. Prenat. Diagn. 19, 282–286.

Farrell, S.A., Winsor, E.J., Markovic, V.D., 1993. Moving satellites and unstable chromosome translocations: Clinical and cytogenetic implications. Am. J. Med. Genet. 46, 715–720.

Feenstra, I., Koolen, D.A., Van der Pas, J., Hamel, B.C., Mieloo, H., Smeets, D.F., Van Ravenswaaij, C.M., 2006 Sep-Oct. Cryptic duplication of the distal segment of 22q due to a translocation (21;22): Three case reports and a review of the literature. Eur. J. Med. Genet. 49 (5), 384–395.

Ferguson-Smith, M.A., Ferguson-Smith, M.E., Ellis, P.M., Dickson, M., 1962. The sites and relative frequencies of secondary constrictions in human somatic chromosomes. Cytogenetics 1, 325–343.

Ferguson-Smith, M.A., 1974. Autosomal polymorphisms. Birth Defects Orig. Artic. Ser. 10, 19–29.

Ferguson-Smith, M.A., Ellis, P.M., Mutchinick, O., Glen, K.P., Côté, G.B., Edwards, J.H., 1975. Centromeric linkage. Cytogenet. Cell Genet. 14, 300–307.

Fernández, J.L., Pereira, S., Campos, A., Goyanes, V., 1994. Assessment of Yqh translocations. J. Med. Genet. 31, 978–979.

Fernández, J.L., Campos, A., López-Fernández, C., Gosálvez, J., Goyanes, V., 1995. Difference in constitutive heterochromatin behaviour between human amniocytes and lymphocytes detected by a sequential in situ exonuclease III digestion-random primer extension procedure. J. Med. Genet. 32, 32–35.

Fernández, J.L., Vázquez-Gundín, F., Rivero, M.T., Goyanes, V., Gosálvez, J., 2001. Evidence of abundant constitutive alkali-labile sites in human 5 bp classical satellite DNA loci by DBD-FISH. Mutat. Res. 473, 163–168.

Fetni, R., Richer, C.L., Malfoy, B., Dutrillaux, B., Lemieux, N., 1997. Cytologic characterization of two distinct alpha satellite DNA domains on human chromosome 7, using double-labeling hybridizations in fluorescence and electron microscopy on a melanoma cell line. Cancer Genet. Cytogenet. 96, 17–22.

Fickelscher, I., Liehr, T., Watts, K., Bryant, V., Barber, J.C., Heidemann, S., Siebert, R., Hertz, J.M., Tumer, Z., Simon Thomas, N., 2007. The variant inv(2)(p11.2q13) is a genuinely recurrent rearrangement but displays some breakpoint heterogeneity. Am. J. Hum. Genet. 81, 847–856.

Filges, I., Röthlisberger, B., Noppen, C., Boesch, N., Wenzel, F., Necker, J., Binkert, F., Huber, A.R., Heinimann, K., Miny, P., 2009. Familial 14.5 Mb interstitial deletion 13q21.1-13q21.33: Clinical and array-CGH study of a benign phenotype in a three-generation family. Am. J. Med. Genet. A 149A, 237–241.

Fineman, R.M., Issa, B., Weinblatt, V., 1989. Prenatal diagnosis of a large heteromorphic region in a chromosome 5: Implications for genetic counseling. Am. J. Med. Genet. 32, 498–499.

Finelli, P., Antonacci, R., Marzella, R., Lonoce, A., Archidiacono, N., Rocchi, M., 1996. Structural organization of multiple alphoid subsets coexisting on human chromosomes 1, 4, 5, 7, 9, 15, 18 and 19. Genomics 38, 325–330.

Finelli, P., Sirchia, S.M., Masciadri, M., Crippa, M., Recalcati, M.P., Rusconi, D., Giardino, D., Monti, L., Cogliati, F., Faravelli, F., Natacci, F., Zoccante, L., Bernardina, B.D., Russo, S., Larizza, L., 2012. Juxtaposition of heterochromatic and euchromatic regions by chromosomal translocation mediates a heterochromatic long-range position effect associated with a severe neurological phenotype. Mol. Cytogenet. 5, 16.

Forabosco, A., Giovannelli, G., Marzona, L., Canè, V., 1976. Trisomy 4p due to translocation t (4p-,22p+). Familial findings in 4 generations. Minerva. Pediatr. 28, 743–751.

Friedrich, U., 1979. C-heteromorphism in chromosome no. 6. Clin. Genet. 16, 295.

Friedrich, U., 1985. Centromere heteromorphism in chromosome 19. Clin. Genet. 28, 358–359.

Friedrich, U., Caprani, M., Niebuhr, E., Therkelsen, A.J., Jørgensen, A.L., 1996. Extreme variant of the short arm of chromosome 15. Hum. Genet. 97, 710–713.

Fryns, J.P., Logghe, N., van Eygen, M., van den Berghe, H., 1979. 18q- syndrome in mother and daughter. Eur. J. Pediatr. 130, 189–192.

Fryns, J.P., Parloir, C., Van den Berghe, H., 1979. Partial trisomy 17q. Karyotype: 46,XY,der(21),t(17;21)(q22;p13). Hum. Genet. 49, 361–364.

Fryns, J.P., Kleczkowska, A., Smeets, E., van den Berghe, H., 1988. A new centromeric heteromorphism in the short arm of chromosome 20. J. Med. Genet. 25, 636–637.

Feuk, L., Carson, A.R., Scherer, S.W., 2006. Structural variation in the human genome. Nat. Rev. Genet. 7, 85–97.

Fu, S.M., 1989. Molecular cytogenetic study of a 21p+ variant in a family with Crouzon syndrome. Zhongguo Yi Xue Ke Xue Yuan Xue Bao 11, 165–169.

Furbetta, M., Rosi, G., Cossu, P., Cao, A., 1975. A case of trisomy of the short arms of chromosome no. 4 with translocation t(4p 21p; 4q 21q) in the mother. Humangenetik 26, 87–91.

Fusco, F., Paciolla, M., Chen, E., Li, X., Genesio, R., Conti, A., Jones, J., Poeta, L., Lioi, M.B., Ursini, M.V., Miano, M.G., 2011. Genetic and molecular analysis of a new unbalanced X;18 rearrangement: Localization of the diminished ovarian reserve disease locus in the distal Xq POF1 region. Hum. Reprod. 26, 3186–3196.

Gaál, M., Tóth, A., Bösze, P., László, J., 1984. 46,X,i(Xq)/45,X mosaicism with gonadal dysgenesis associated with 21p-. Clin. Genet. 25, 79–83.

Gardner, H.A., Wood, E.M., 1979. Variation in chromosome 19. J. Med. Genet. 16, 79–80.

Gardner, R.J.M., Sutherland, G.R., 2004. Chromosome abnormalities and genetic counseling. Oxford Monographs on Medical Genetcis. Oxford University Press.

Gardner, R.J.M., Sutherland, G.R., Shaffer, L.G., 2012. Chromosome abnormalities and genetic counseling. Oxford Monographs on Medical Genetcis. Oxford University Press.

Gar'kavtsev, I.V., Tsvetkova, T.G., Egolina, N.A., Mkhitarova, E.V., Gudkov, A.V., 1986. Molecular and cytogenetic approach to the mapping of methylated polymorphic restriction fragments of human rRNA genes. Biull. Eksp. Biol. Med. 102, 330–331.

Ge, Y., Wagner, M.J., Siciliano, M., Wells, D.E., 1992. Sequence, higher order repeat structure, and long-range organization of alpha satellite DNA specific to human chromosome 8. Genomics 13, 585–593.

Gebauer, H.J., Scheil, H.G., Röhrborn, G., 1988. Partial deletion of the short arm of chromosome 13 as an indication of paternity in forensic assessment. Z Rechtsmed. 99, 249–251.

Genest, P., 1972. An eleven-generation satellited Y-chromosome. Lancet I 1073.

Genest, P., Bouchard, M., Bouchard, J., 1967. A satellited human Y-chromosome. Lancet I, 1279–1280.

Geraedts, J.P., Pearson, P.L., 1974. Fluorescent chromosome polymorphisms: Frequencies and segregations in a Dutch population. Clin. Genet. 6, 247–257.

Ghoumid, J., Andrieux, J., Sablonnière, B., Odent, S., Philippe, N., Zanlonghi, X., Saugier-Veber, P., Bardyn, T., Manouvrier-Hanu, S., Holder-Espinasse, M., 2011. Duplication at chromosome 2q31.1-q31.2 in a family presenting syndactyly and nystagmus. Eur. J. Hum. Genet. 19, 1198–1201.

Gilgenkrantz, S., Vigneron, J., Peter, M.O., Dufier, J.L., Teboul, M., Chery, M., Keyeux, G., Lefranc, M.P., 1990. Distal trisomy 14q. I. Clinical and cytogenetical studies. Hum. Genet. 85, 612–616.

Gilling, M., Dullinger, J.S., Gesk, S., Metzke-Heidemann, S., Siebert, R., Meyer, T., Brondum-Nielsen, K., Tommerup, N., Ropers, H.H., Tumer, Z., Kalscheuer, V.M., Thomas, N.S., 2006. Breakpoint cloning and haplotype analysis indicate a single origin of the common inv(10)(p11.2q21.2) mutation among northern Europeans. Am. J. Hum. Genet. 78, 878–883.

Giraldo, A., Martínez, I., Guzmán, M., Silva, E., 1981. A family with a satellited Yq chromosome. Hum. Genet. 57, 99–100.

Girirajan, S., Rosenfeld, J.A., Cooper, G.M., Antonacci, F., Siswara, P., Itsara, A., Vives, L., Walsh, T., McCarthy, S.E., Baker, C., Mefford, H.C., Kidd, J.M., Browning, S.R., Browning, B.L., Dickel, D.E., Levy, D.L., Ballif, B.C., Platky, K., Farber, D.M., Gowans, G.C., Wetherbee, J.J., Asamoah, A., Weaver, D.D., Mark, P.R., Dickerson, J., Garg, B.P., Ellingwood, S.A., Smith, R., Banks, V.C., Smith, W., McDonald, M.T., Hoo, J.J., French, B.N., Hudson, C., Johnson, J.P., Ozmore, J.R., Moeschler, J.B., Surti, U., Escobar, L.F., El-Khechen, D., Gorski, J.L., Kussmann, J., Salbert, B., Lacassie, Y., Biser, A., McDonald-McGinn, D.M., Zackai, E.H., Deardorff, M.A., Shaikh, T.H., Haan, E., Friend, K.L., Fichera, M., Romano, C., Gécz, J., DeLisi, L.E., Sebat, J., King, M.C., Shaffer, L.G., Eichler, E.E., 2010. A recurrent 16p12.1 microdeletion supports a two-hit model for severe developmental delay. Nat. Genet. 42, 203–209.

Girirajan, S., Campbell, C.D., Eichler, E.E., 2011. Human copy number variation and complex genetic disease. Annu. Rev. Genet. 45, 203–226.

Giussani, U., Facchinetti, B., Cassina, G., Zuffardi, O., 1996. Mitotic recombination among acrocentric chromosomes' short arms. Ann. Hum. Genet. 60, 91–97.

Glancy, M., Barnicoat, A., Vijeratnam, R., de Souza, S., Gilmore, J., Huang, S., Maloney, V.K., Thomas, N.S., Bunyan, D.J., Jackson, A., Barber, J.C., 2009. Transmitted duplication of 8p23.1-8p23.2 associated with speech delay, autism and learning difficulties. Eur. J. Hum. Genet. 17, 37–43.

Göhring, I., Blümlein, H.M., Hoyer, J., Ekici, A.B., Trautmann, U., Rauch, A., 2008. 6.7 Mb interstitial duplication in chromosome band 11q24.2q25 associated with infertility, minor dysmorphic features and normal psychomotor development. Eur. J. Med. Genet. 51, 666–671.

González García, J.R., Garcés Ruíz, O.M., Delgado Lamas, J.L., Ramírez-Dueñas, M.L., 1997. Two different Philadelphia chromosomes in a cell line from an AML-M0 patient. Cancer Genet. Cytogenet. 98, 111–114.

Goumy, C., Gouas, L., Tchirkov, A., Roucaute, T., Giollant, M., Veronèse, L., Francannet, C., Vago, P., 2008. Familial deletion 11q14.3-q22.1 without apparent phenotypic consequences: A haplosufficient 8.5 Mb region. Am. J. Med. Genet. A 146A, 2668–2672.

Goumy, C., Kemeny, S., Eymard-Pierre, E., Richard, C., Gouas, L., Combes, P., Gay-Bellile, M., Gallot, D., Tchirkov, A., Vago, P., 2011. Prenatal diagnosis of a rare de novo centromeric chromosome 6 variant. Gene 490, 15–17.

Gourzis, P., Skokou, M., Polychronopoulos, P., Soubasi, E., Triantaphyllidou, I.E., Aravidis, C., Sarela, A.I., Kosmaidou, Z., 2012. Frontotemporal dementia, manifested as schizophrenia, with decreased heterochromatin on chromosome 1. Case Rep. Psychiatry. 2012, 937518.

Gouw, W.L., Anders, G.J., ten Kate, L.P., de Groot, C.J., 1972. Paternal transmission of a B-D translocation, t(4p-; 14p + or 15p+), resulting in a partial 4p trisomy. Humangenetik 16, 251–259.

Greig, G.M., Willard, H.F., 1992. Beta satellite DNA: Characterization and localization of two subfamilies from the distal and proximal short arms of the human acrocentric chromosomes. Genomics 12, 573–580.

Greig, G.M., England, S.B., Bedford, H.M., Willard, H.F., 1989. Chromosome-specific alpha satellite DNA from the centromere of human chromosome 16. Am. J. Hum. Genet. 45, 862–872.

Griffiths, M.J., Miller, P.R., Stibbe, H.M., 1996. A false-positive diagnosis of Turner syndrome by amniocentesis. Prenat. Diagn. 16, 463–466.

Grunau, C., Buard, J., Brun, M.E., De Sario, A., 2006. Mapping of the juxtacentromeric heterochromatin-euchromatin frontier of human chromosome 21. Genome Res. 16, 1198–1207.

Gu, W., Zhang, F., Lupski, J.R., 2008. Mechanisms for human genomic rearrangements. Pathogenetics 1, 4.

Guanti, G., Mollica, G., Polimeno, L., Maritato, F., 1976. rDNA and acrocentric chromosomes in man. I. rDNA levels in a subject carrier of a 8p/13p balanced translocation and in his unbalanced son. Hum. Genet. 33, 103–107.

Guissani, U., Facchinetti, B., Cassina, G., Zuffardi, O., 1996. Mitotic recombination among acrocentric chromosomes' short arms. Ann. Hum. Genet. 60, 91–97.

Gunel, M., Cavkaytar, S., Ceylaner, G., Batioglu, S., 2008. Azoospermia and cryptorchidism in a male with a de novo reciprocal t(Y;16) translocation. Genet. Couns. 19, 277–280.

Guo, Q.S., Qin, S.Y., Zhou, S.F., He, L., Ma, D., Zhang, Y.P., Xiong, Y., Peng, T., Cheng, Y., Li, X.T., 2009. Unbalanced translocation in an adult patient with premature ovarian failure and mental retardation detected by spectral karyotyping and array-comparative genomic hybridization. Eur. J. Clin. Invest. 39, 729–737.

Gustashaw, K.M., Zurcher, V., Dickerman, L.H., Stallard, R., Willard, H.F., 1994. Partial X chromosome trisomy with functional disomy of Xp due to failure of X inactivation. Am. J. Med. Genet. 53, 39–45.

Guttenbach, M., Nassar, N., Feichtinger, W., Steinlein, C., Nanda, I., Wanner, G., Kerem, B., Schmid, M., 1998. An interstitial nucleolus organizer region in the long arm of human chromosome 7: Cytogenetic characterization and familial segregation. Cytogenet. Cell Genet. 80, 104–112.

Guttenbach, M., Haaf, T., Steinlein, C., Caesar, J., Schinzel, A., Schmid, M., 1999. Ectopic NORs on human chromosomes 4qter and 8q11: Rare chromosomal variants detected in two families. J. Med. Genet. 36, 339–342.

Haaf, T., Feichtinger, W., Guttenbach, M., Sanchez, L., Müller, C.R., Schmid, M., 1989. Berenil-induced undercondensation in human heterochromatin. Cytogenet. Cell Genet. 50, 27–33.

Haaf, T., Willard, H.F., 1992. Organization, polymorphism, and molecular cytogenetics of chromosome-specific alpha-satellite DNA from the centromere of chromosome 2. Genomics 13, 122–128.

Habibian, R., Hajianpour, M.J., Shaffer, L.G., Niedenard, L., Hajianpour, A.k., 1994. Geneotype-phenotype correlation in satellited 1p chromosome: Importance of fluorescence in situ hybridization (FISH) applicatiomns. Am. J. Hum. Genet. 55 (Suppl.), A106.

Hall, L.E., Mitchell, S.E., O'Neill, R.J., 2012. Pericentric and centromeric transcription: A perfect balance required. Chromosome Res. 20, 535–546.

Hamerton, J.L., Klinger, H.P., Mutton, D.E., Lang, E.M., 1963. The somatoc chromosomes of the Hominoidea. Cytogenetics 25, 240–263.

Hamerton, J.L., Ray, M., Abbott, J., Williamson, C., Ducasse, G.C., 1972. Chromosome studies in a neonatal population. Can. Med. Assoc. J. 106, 776–779.

Hamerton, J.L., Canning, N., Ray, M., Smith, S., 1975. A cytogenetic survey of 14,069 newborn infants. I. Incidence of chromosome abnormalities. Clin. Genet. 8, 223–243.

Hancke, I., Miller, K., 1985. Familial occurrence of a pseudodicentric chromosome 21. J. Med. Genet. 22, 155–156.

Hansson, K.B., Gijsbers, A.C., Oostdijk, W., Rehbock, J.J., de Snoo, F., Ruivenkamp, C.A., Kant, S.G., 2012. Molecular and clinical characterization of patients with a ring chromosome 11. Eur. J. Med. Genet. 55, 708–714.

Hansson, K., Szuhai, K., Knijnenburg, J., van Haeringen, A., de Pater, J., 2007. Interstitial deletion of 6q without phenotypic effect. Am. J. Med. Genet. A 143A, 1354–1357.

Hansmann, I., 1976. Structural variability of human chromosome 9 in relation to its evolution. Hum. Genet. 31, 247–262.

Hansteen, I.L., Schirmer, L., Hestetun, S., 1978. Trisomy 12p syndrome. Evaluation of a family with a t(12;21) (p12.1;p11) translocation with unbalanced offspring. Clin. Genet. 13, 339–349.

Harada, N., Takano, J., Kondoh, T., Ohashi, H., Hasegawa, T., Sugawara, H., Ida, T., Yoshiura, K., Ohta, T., Kishino, T., Kajii, T., Niikawa, N., Matsumoto, N., 2002. Duplication of 8p23.2: A benign cytogenetic variant? Am. J. Med. Genet. 111, 285–288.

Hardas, B.D., Zhang, J., Trent, J.M., Elder, J.T., 1994. Direct evidence for homologous sequences on the paracentric regions of human chromosome 1. Genomics 21, 359–363.

Harris, P., Cooke, A., Boyd, E., Young, B.D., Ferguson-Smith, M.A., 1987. The potential of family flow karyotyping for the detection of chromosome abnormalities. Hum. Genet. 76, 129–133.

Hengstschläger, M., Prusa, A., Repa, C., Deutinger, J., Pollak, A., Bernaschek, G., 2005. Subtelomeric rearrangements as neutral genomic polymorphisms. Am. J. Med. Genet. A 133A, 48–52.

Henn, B.M., Cavalli-Sforza, L.L., Feldman, M.W., 2012. The great human expansion. Proc. Natl. Acad. Sci. U. S. A. 109, 17758–17764.

Henson, K.E., Hines, K.A., Weaver, D.D., Torres, W.M., Verbrugge, J., Stone, K., Vance, G.H., 2012. Duplication of 18q21.32-q22.3 identified in a stillborn and two relatives with minimal dysmorphic features. Am. J. Med. Genet. A 158A, 1788–1792.

Higgins, M.J., Wang, H.S., Shtromas, I., Haliotis, T., Roder, J.C., Holden, J.J., White, B.N., 1985. Organization of a repetitive human 1.8 kb KpnI sequence localized in the heterochromatin of chromosome 15. Chromosoma 93, 77–86.

Horvath, J.E., Viggiano, L., Loftus, B.J., Adams, M.D., Archidiacono, N., Rocchi, M., Eichler, E.E., 2000. Molecular structure and evolution of an alpha satellite/non-alpha satellite junction at 16p11. Hum. Mol. Genet. 9, 113–123.

Horvath, J.E., Bailey, J.A., Locke, D.P., Eichler, E.E., 2001. Lessons from the human genome: Transitions between euchromatin and heterochromatin. Hum. Mol. Genet. 10, 2215–2223.

Horvath, J.E., Gulden, C.L., Bailey, J.A., Yohn, C., McPherson, J.D., Prescott, A., Roe, B.A., de Jong, P.J., Ventura, M., Misceo, D., Archidiacono, N., Zhao, S., Schwartz, S., Rocchi, M., Eichler, E.E., 2003. Using a pericentromeric interspersed repeat to recapitulate the phylogeny and expansion of human centromeric segmental duplications. Mol. Biol. Evol. 20, 1463–1479.

Hoshi, N., Fujita, M., Mikuni, M., Fujino, T., Okuyama, K., Handa, Y., Yamada, H., Sagawa, T., Hareyama, H., Nakahori, Y., Fujieda, K., Kant, J.A., Nagashima, K., Fujimoto, S., 1998. Seminoma in a postmenopausal woman with a Y;15 translocation in peripheral blood lymphocytes and a t(Y;15)/45, X Turner mosaic pattern in skin fibroblasts. J. Med. Genet. 35, 852–856.

Horsley, S.W., Daniels, R.J., Anguita, E., Raynham, H.A., Peden, J.F., Villegas, A., Vickers, M.A., Green, S., Waye, J.S., Chui, D.H., Ayyub, H., MacCarthy, A.B., Buckle, V.J., Gibbons, R.J., Kearney, L., Higgs, D.R., 2001. Monosomy for the most telomeric, gene-rich region of the short arm of human chromosome 16 causes minimal phenotypic effects. Eur. J. Hum. Genet. 9, 217–225.

Hou, J.W., Wang, T.R., 1995. Molecular cytogenetic studies of duplication 9q32–>q34.3 inserted into 9q13. Clin. Genet. 48, 148–150.

Howell, W.M., Howard-Peebles, P.N., Block, B.M., Stoddard, G.R., 1978. Silver stain reveals nucleolus organizer regions on a satellited Yq chromosome. Hum. Genet. 42, 245–250.

Hsu, L.Y., Benn, P.A., Tannenbaum, H.L., Perlis, T.E., Carlson, A.D., 1987. Chromosomal polymorphisms of 1, 9, 16, and Y in 4 major ethnic groups: A large prenatal study. Am. J. Med. Genet. 26, 95–101.

Hulsebos, T., Schonk, D., van Dalen, I., Coerwinkel-Driessen, M., Schepens, J., Ropers, H.H., Wieringa, B., 1988. Isolation and characterization of alphoid DNA sequences specific for the pericentric regions of chromosomes 4, 5, 9, and 19. Cytogenet. Cell Genet. 47, 144–148.

Ibraimov, A.I., Mirrakhimov, M.M., Axenrod, E.I., 1986. Human chromosomal polymorphism. VIII. Chromosomal Q polymorphism in the Yakut of Eastern Siberia. Hum. Genet. 73, 147–150.

Iourov, I.Y., Soloviev, I.V., Vorsanova, S.G., Monakhov, V.V., Yurov, Y.B., 2005. An approach for quantitative assessment of fluorescence in situ hybridization (FISH) signals for applied human molecular cytogenetics. J. Histochem. Cytochem. 53, 401–408.

Iurov Iu, B., Selivanova, E.A., Deriagin, G.V., 1991. Human alpha-satellite DNA specific to chromosomes 13 and 21: Use for the analysis of polymorphism of acrocentric chromosomes and the origin of the additional chromosome 21 in Down's syndrome. Genetika 27, 1637–1647.

ISCN, 1978. An international system for human cytogenetic nomenclature (1978). In: Lindsten, J.E., Klinger, H.P., Hamerton, J.L. (Eds.), Karger, Basel, 1978.

ISCN, 1985. An international system for human cytogenetic nomenclature (1985). In: Harnden, D.G., Klinger, H.P. (Eds.), Karger, Basel, 1985.

ISCN, 2005. An international system for human cytogenetic nomenclature (2005). In: Shaffer, L.G., Tommerup, N. (Eds.), Karger, Basel, 2005.

ISCN, 2009. An international system for human cytogenetic nomenclature. In: Shaffer, L.G., Slovak, M.L., Campbell, L.J. (Eds.), Karger, Basel, 2009.

ISCN, 2013. An international system for human cytogenetic nomenclature. In: Shaffer, L.G., McGowan-Jordan, J., Schmid, M. (Eds.), Karger, Basel, 2013.

Izakovic, V., 1984. Monosomy X associated with fra(17p12) and 22p-. Clin. Genet. 26, 165–166.

Jabs, E.W., Carpenter, N., 1988. Molecular cytogenetic evidence for amplification of chromosome-specific alphoid sequences at enlarged C-bands on chromosome 6. Am. J. Hum. Genet. 43, 69–74.

Jackson, M.S., Slijepcevic, P., Ponder, B.A., 1993. The organisation of repetitive sequences in the pericentromeric region of human chromosome 10. Nucleic Acids Res. 21, 5865–5874.

Jacobs, P.A., 1992. The chromosome complement of human gametes. Oxf. Rev. Reprod. Biol. 14, 47–72.

Jacobsen, P., Mikkelsen, M., 1968. Chromosome 18 abnormalities in a family with a translocation t(18p–, 21p+). J. Ment. Defic. Res. 12, 144–161.

Jalal, S.M., Ketterling, R.P., 2004. Euchromatic variants. In: Wyandt, H.E., Tonk, V.S. (Eds.), Atlas of Human Chromosome Heteromorphisms. Kluwer, pp. 75–86.

Jalal, S.M., Clark, R.W., Hsu, T.C., Pathak, S., 1974. Cytological differentiation of constitutive heterochromatin. Chromosoma 48, 391–403.

Jenkyn, D.J., Whitehead, R.H., House, A.K., Maley, M.A., 1987. Single chromosome defect, partial trisomy 1q, in a colon cancer cell line. Cancer Genet. Cytogenet. 27, 357–360.

Jobanputra, V., Sebat, J., Troge, J., Chung, W., Anyane-Yeboa, K., Wigler, M., Warburton, D., 2005. Application of ROMA (representational oligonucleotide microarray analysis) to patients with cytogenetic rearrangements. Genet. Med. 7, 111–118.

Jobling, M.A., 2008. Copy number variation on the human Y chromosome. Cytogenet. Genome Res. 123, 253–362.

Jørgensen, A.L., 1997. Alphoid repetitive DNA in human chromosomes. Dan. Med. Bull. 44, 522–534.

Kallioniemi, A., Kallioniemi, O.P., Sudar, D., Rutovitz, D., Gray, J.W., Waldman, F., Pinkel, D., 1992. Comparative genomic hybridization for molecular cytogenetic analysis of solid tumors. Science 258, 818–821.

Kamei, T., Lee-Okimoto, S., Sohda, M., Niikawa, N., 1986. A further improved method for identifying heteromorphism of acrocentric chromosomes. Hum. Genet. 73, 368–371.

Kar, B., Prabhakara, K., Murthy, S.K., 1992. Inherited deletion of chromosome (21p-) in a child with congenital malformation and psychomotor retardation. Indian Pediatr. 29, 929–934.

Keyeux, G., Gilgenkrantz, S., Lefranc, G., Lefranc, M.P., 1990. Distal trisomy 14q. II. Molecular study of the 14q32 locus in two cases of chromosome 14 rearrangements with partial duplication. Hum. Genet. 85, 617–622.

Ki, A., Rauen, K.A., Black, L.D., Kostiner, D.R., Sandberg, P.L., Pinkel, D., Albertson, D.G., Norton, M.E., Cotter, P.D., 2003. Ring 21 chromosome and a satellited 1p in the same patient: novel origin for an ectopic NOR. Am. J. Med. Genet. A 120A, 365–369.

Killos, L.D., Lese, C.M., Mills, P.L., Precht, K.S., Stanley, W.S., Ledbetter, D.H., 1997. A satellited 17p with telomere deleted and no apparent clinical consequences. Am. J. Hum. Genet. 61 (Suppl.), A130.

Kim, M.A., 1975. Fluorometrical detection of thymine base differences in complementary strands of satellite DNA in human metaphase chromosomes. Humangenetik 28, 57–63.

Kim, J.W., Park, J.Y., Oh, A.R., Choi, E.Y., Ryu, H.M., Kang, I.S., Koong, M.K., Park, S.Y., 2011. Duplication of intrachromosomal insertion segments 4q32 → q35 confirmed by comparative genomic hybridization and fluorescent in situ hybridization. Clin. Exp. Reprod. Med. 38, 238–241.

Kleczkowska, A., Fryns, J.P., Jaeken, J., Van den Berghe, H., 1988. Complex chromosomal rearrangement involving chromosomes 11, 13 and 21. Ann. Genet. 31, 126–128.

Kleinjan, D.J., van Heyningen, V., 1998. Position effect in human genetic disease. Hum. Mol. Genet. 7, 1611–1618.

Knight, S.J., Flint, J., 2000. Perfect endings: A review of subtelomeric probes and their use in clinical diagnosis. J. Med. Genet. 37, 401–409.

Knijnenburg, J., van Bever, Y., Hulsman, L.O., van Kempen, C.A., Bolman, G.M., van Loon, R.L., Beverloo, H.B., van Zutven, L.J., 2012. A 600 kb triplication in the cat eye syndrome critical region causes anorectal, renal and preauricular anomalies in a three-generation family. Eur. J. Hum. Genet. 20, 986–989.

Knuutila, S., Grippenberg, U., 1972. The fluorescence pattern of a human Yq+ chromosome. Hereditas 70, 307–308.

Kosyakova, N., Weise, A., Mrasek, K., Claussen, U., Liehr, T., Nelle, H., 2009. The hierarchically organized splitting of chromosomal bands for all human chromosomes. Mol. Cytogenet. 2, 4.

Kosyakova, N., Grigorian, A., Liehr, T., Manvelyan, M., Simonyan, I., Mkrtchyan, H., Aroutiounian, R., Polityko, A.D., Kulpanovich, A.I., Egorova, T., Jaroshevich, E., Frolova, A., Shorokh, N., Naumchik, I.V., Volleth, M., Schreyer, I., Nelle, H., Stumm, M., Wegner, R.-D., Reising-Ackermann, G., Merkas, M., Brecevic, L., Martin, T., Rodríguez, L., Bhatt, S., Ziegler, M., Kreskowski, K., Weise, A., Sazci, A., Iourov, I., de Bello Cioffi, M., Ergul, E., 2013. Heteromorphic variants of chromosome 9. Mol. Cytogenet. 6, 14.

Kristoffersson, U., 2008. Regulatory issues for genetic testing in clinical practice. Mol. Biotechnol. 40, 113–117.

Kristoffersson, U., Schmidtke, J., Cassiman, J.J., 2010. Quality Issues in Clinical Genetic Services. Springer, Berlin.

Kruminia, A.R., Kroshkina, V.G., Iurov IuB, Aleksandrov IA, Mitkevich, S.P., 1988. Use of a cloned alphoid repetitive sequence of human DNA in studying the polymorphism of heterochromatin regions of chromosomes. Genetika 24, 937–942.

Kuan, L.C., Su, M.T., Kuo, P.L., Kuo, T.C., 2013 Apr. Direct duplication of the Y chromosome with normal phenotype—Incidental finding in two cases. Andrologia 45 (2), 140–144.

Kühl, H., Röttger, S., Heilbronner, H., Enders, H., Schempp, W., 2001. Loss of the Y chromosomal PAR2-region in four familial cases of satellited Y chromosomes (Yqs). Chromosome Res. 9, 215–222.

Kuiper, R.P., Ligtenberg, M.J., Hoogerbrugge, N., Geurts van Kessel, A., 2010. Germline copy number variation and cancer risk. Curr. Opin. Genet. Dev. 20, 282–289.

Kurotaki, N., Shen, J.J., Touyama, M., Kondoh, T., Visser, R., Ozaki, T., Nishimoto, J., Shiihara, T., Uetake, K., Makita, Y., Harada, N., Raskin, S., Brown, C.W., Höglund, P., Okamoto, N., Lupski, J.R., 2005. Phenotypic consequences of genetic variation at hemizygous alleles: Sotos syndrome is a contiguous gene syndrome incorporating coagulation factor twelve (FXII) deficiency. Genet. Med. 7, 479–483.

Kuznetzova, T.V., Trofimova, I.L., Liapunov, M.S., Evdokimenko, E.V., Baranov, V.S., 2012. Selective staining of pericentromeric heterochromatin regions in chromosomes of spontaneously dividing cells with the use of the acridine orange fluorochrome. Genetika 48, 451–456.

Lamb, A.N., Pettanati, M., Hanna, J., Krasikov, N., Neu, R., Rao, N., Weinstein, M., Weiser, J., Estabrooks, L., 1995. Six cases of satellited long arm of chromosome 2 detected during prenatal chromosome diagnosis. Am. J. Hum. Genet. 57 (Suppl.), A282.

Lander, E.S., Linton, L.M., Birren, B., Nusbaum, C., Zody, M.C., Baldwin, J., Devon, K., Dewar, K., Doyle, M., FitzHugh, W., Funke, R., Gage, D., Harris, K., Heaford, A., Howland, J., Kann, L., Lehoczky, J., LeVine, R., McEwan, P., McKernan, K., Meldrim, J., Mesirov, J.P., Miranda, C., Morris, W., Naylor, J., Raymond, C., Rosetti, M., Santos, R., Sheridan, A., Sougnez, C., Stange-Thomann, N., Stojanovic, N., Subramanian, A., Wyman, D., Rogers, J., Sulston, J., Ainscough, R., Beck, S., Bentley, D., Burton, J., Clee, C., Carter, N., Coulson, A., Deadman, R., Deloukas, P., Dunham, A., Dunham, I., Durbin, R., French, L., Grafham, D., Gregory, S., Hubbard, T., Humphray, S., Hunt, A., Jones, M., Lloyd, C., McMurray, A., Matthews, L., Mercer, S., Milne, S., Mullikin, J.C., Mungall, A., Plumb, R., Ross, M., Shownkeen, R., Sims, S., Waterston, R.H., Wilson, R.K., Hillier, L.W., McPherson, J.D., Marra, M.A., Mardis, E.R., Fulton, L.A., Chinwalla, A.T., Pepin, K.H., Gish, W.R., Chissoe, S.L., Wendl, M.C., Delehaunty, K.D., Miner, T.L., Delehaunty, A., Kramer, J.B., Cook, L.L., Fulton, R.S., Johnson, D.L., Minx, P.J., Clifton, S.W., Hawkins, T., Branscomb, E., Predki, P., Richardson, P., Wenning, S., Slezak, T., Doggett, N., Cheng, J.F., Olsen, A., Lucas, S., Elkin, C., Uberbacher, E., Frazier, M., Gibbs, R.A., Muzny, D.M., Scherer, S.E., Bouck, J.B., Sodergren, E.J., Worley, K.C., Rives, C.M., Gorrell, J.H., Metzker, M.L., Naylor, S.L., Kucherlapati, R.S., Nelson, D.L., Weinstock, G.M., Sakaki, Y., Fujiyama, A., Hattori, M., Yada, T., Toyoda, A., Itoh, T., Kawagoe, C., Watanabe, H., Totoki, Y., Taylor, T., Weissenbach, J., Heilig, R., Saurin, W., Artiguenave, F., Brottier, P., Bruls, T., Pelletier, E., Robert, C., Wincker, P., Smith, D.R., Doucette-Stamm, L., Rubenfield, M., Weinstock, K., Lee, H.M., Dubois, J., Rosenthal, A., Platzer, M., Nyakatura, G., Taudien, S., Rump, A., Yang, H., Yu, J., Wang, J.,

Huang, G., Gu, J., Hood, L., Rowen, L., Madan, A., Qin, S., Davis, R.W., Federspiel, N.A., Abola, A.P., Proctor, M.J., Myers, R.M., Schmutz, J., Dickson, M., Grimwood, J., Cox, D.R., Olson, M.V., Kaul, R., Raymond, C., Shimizu, N., Kawasaki, K., Minoshima, S., Evans, G.A., Athanasiou, M., Schultz, R., Roe, B.A., Chen, F., Pan, H., Ramser, J., Lehrach, H., Reinhardt, R., McCombie, W.R., de la Bastide, M., Dedhia, N., Blöcker, H., Hornischer, K., Nordsiek, G., Agarwala, R., Aravind, L., Bailey, J.A., Bateman, A., Batzoglou, S., Birney, E., Bork, P., Brown, D.G., Burge, C.B., Cerutti, L., Chen, H.C., Church, D., Clamp, M., Copley, R.R., Doerks, T., Eddy, S.R., Eichler, E.E., Furey, T.S., Galagan, J., Gilbert, J.G., Harmon, C., Hayashizaki, Y., Haussler, D., Hermjakob, H., Hokamp, K., Jang, W., Johnson, L.S., Jones, T.A., Kasif, S., Kaspryzk, A., Kennedy, S., Kent, W.J., Kitts, P., Koonin, E.V., Korf, I., Kulp, D., Lancet, D., Lowe, T.M., McLysaght, A., Mikkelsen, T., Moran, J.V., Mulder, N., Pollara, V.J., Ponting, C.P., Schuler, G., Schultz, J., Slater, G., Smit, A.F., Stupka, E., Szustakowski, J., Thierry-Mieg, D., Thierry-Mieg, J., Wagner, L., Wallis, J., Wheeler, R., Williams, A., Wolf, Y.I., Wolfe, K.H., Yang, S.P., Yeh, R.F., Collins, F., Guyer, M.S., Peterson, J., Felsenfeld, A., Wetterstrand, K.A., Patrinos, A., Morgan, M.J., de Jong, P., Catanese, J.J., Osoegawa, K., Shizuya, H., Choi, S., Chen, Y.J., 2001. International Human Genome Sequencing Consortium. Initial sequencing and analysis of the human genome. Nature 409, 860–921.

Langer, S., Fauth, C., Rocchi, M., Murken, J., Speicher, M.R., 2001. AcroM fluorescent in situ hybridization analyses of marker chromosomes. Hum. Genet. 109, 152–158.

Lapidot-Lifson, Y., Lebo, R.V., Flandermeyer, R.R., Chung, J.H., Golbus, M.S., 1996. Rapid aneuploid diagnosis of high-risk fetuses by fluorescence in situ hybridization. Am. J. Obstet. Gynecol. 174, 886–890.

Lau, Y.F., Wertelecki, W., Pfeiffer, R.A., Arrighi, F.E., 1979. Cytological analyses of 14p+ variant by means of N-banding and combinations of silver staining and chromosome bandings. Hum. Genet. 46, 75–82.

Laurent, A.M., Li, M., Sherman, S., Roizès, G., Buard, J., 2003. Recombination across the centromere of disjoined and non-disjoined chromosome 21. Hum. Mol. Genet. 12, 2229–2239.

LeChien, K.A., McPherson, E., Estop, A.M., 1994. Duplication 20p identified via fluorescent in situ hybridization. Am. J. Med. Genet. 50, 187–189.

Lee, C., Wevrick, R., Fisher, R.B., Ferguson-Smith, M.A., Lin, C.C., 1997. Human centromeric DNAs. Hum. Genet. 100, 291–304.

Lee, C., Critcher, R., Zhang, J.G., Mills, W., Farr, C.J., 2000. Distribution of gamma satellite DNA on the human X and Y chromosomes suggests that it is not required for mitotic centromere function. Chromosoma. 109, 381–389.

Lee, M.H., Park, S.Y., Kim, Y.M., Kim, J.M., Yoo, K.J., Lee, H.H., Ryu, H.M., 2005. Molecular cytogenetic characterization of ring chromosome 4 in a female having a chromosomally normal child. Cytogenet. Genome Res. 111, 175–178.

Lee, C., Iafrate, A.J., Brothman, A.R., 2007. Copy number variations and clinical cytogenetic diagnosis of constitutional disorders. Nat. Genet. 39, S48–S54.

Lejeune, J., Gautier, M., Turpin, R., 1959. Étude des chromosomes somatiques de neuf enfants mongoliens. C. R. Acad. Sci. Paris 248, 1721–1722, and Bull Acad Med 1959; 143:11–12; 256–265.

Lemmers, R.J., Tawil, R., Petek, L.M., Balog, J., Block, G.J., Santen, G.W., Amell, A.M., van der Vliet, P.J., Almomani, R., Straasheijm, K.R., Krom, Y.D., Klooster, R., Sun, Y., den Dunnen, J.T., Helmer, Q., Donlin-Smith, C.M., Padberg, G.W., van Engelen, B.G., de Greef, J.C., Aartsma-Rus, A.M., Frants, R.R., de Visser, M., Desnuelle, C., Sacconi, S., Filippova, G.N., Bakker, B., Bamshad, M.J., Tapscott, S.J., Miller, D.G., van der Maarel, S.M., 2012. Digenic inheritance of an SMCHD1 mutation and an FSHD-permissive D4Z4 allele causes facioscapulohumeral muscular dystrophy type 2. Nat. Genet. 44, 1370–1374.

Levy, B., Dunn, T.M., Hirschhorn, K., Kardon, N., 2000. Jumping translocations in spontaneous abortions. Cytogenet. Cell Genet. 88, 25–29.

Liehr, T., 2009. Fluorescence in situ Hybridization (FISH)—Application Guide. Springer, Berlin.

Liehr, T., 2010. Cytogenetic contribution to uniparental disomy (UPD). Mol. Cytogenet. 3, 8.

Liehr, T., 2012. Small Supernumerary Marker Chromosomes (sSMC) A Guide for Human Geneticists and Clinicians; With contributions by UNIQUE (The Rare Chromosome Disorder Support Group). Springer, Heidelberg.

Liehr, T., 2013. http://www.fish.uniklinikum-jena.de/sSMC.html.

Liehr, T., Claussen, U., 2002. Current developments in human molecular cytogenetic techniques. Cur. Mol. Med. 2, 269–284.

Liehr, T., Pfeiffer, R.A., Trautmann, U., 1992. Typical and partial cat-eye-syndrome: Identification of the marker chromosome by FISH. Clin. Genet. 42, 91–96.

Liehr, T., Park, O., Feuerstein, B., Gebhart, E., Rautenstrauss, B., 1997. The peripheral myelin protein 22 kDa (PMP22) gene is amplified in cell lines derived from glioma and osteogenic sarcoma. Int. J. Oncol. 10, 915–919.

Liehr, T., Pfeiffer, R.A., Trautmann, U., Gebhart, E., 1998. Centromeric alphoid DNA heteromorphisms of chromosome 22 revealed by FISH-technique. Clin. Genet. 53, 231–232.

Liehr, T., Starke, H., Beensen, V., Kähler, C., Harbich, M., Brude, E., Ziegler, M., Claussen, U., 1999. Translocation trisomy dup(21q) and free trisomy 21 can be distinguished by interphase-FISH. Int. J. Mol. Med. 3, 11–14.

Liehr, T., Beensen, V., Hauschild, R., Ziegler, M., Hartmann, I., Starke, H., Heller, A., Kähler, C., Schmidt, M., Reiber, W., Hesse, M., Claussen, U., 2001. Pitfalls of rapid prenatal diagnosis using the interphase nucleus. Prenat. Diagn. 21, 419–421.

Liehr, T., Schreyer, I., Neumann, A., Beensen, V., Ziegler, M., Hartmann, I., Starke, H., Heller, A., Nietzel, A., Claussen, U., 2002. Two more possible pitfalls of rapid prenatal diagnostics using interphase nuclei. Prenat. Diagn. 22, 497–499.

Liehr, T., Ziegler, M., Starke, H., Heller, A., Kuechler, A., Kittner, G., Beensen, V., Seidel, J., Hässler, H., Müsebeck, J., Claussen, U., 2003. Conspicuous GTG-banding results of the centromere-near region can be caused by alphoid DNA heteromorphism. Clin. Genet. 64, 166–167.

Liehr, T., Mrasek, K., Weise, A., Dufke, A., Rodríguez, L., Martínez Guardia, N., Sanchís, A., Vermeesch, J.R., Ramel, C., Polityko, A., Haas, O.A., Anderson, J., Claussen, U., von Eggeling, F., Starke, H., 2006. Small supernumerary marker chromosomes—Progress towards a genotype-phenotype correlation. Cytogenet. Genome Res. 112, 23–34.

Liehr, T., Mrasek, K., Hinreiner, S., Reich, D., Ewers, E., Bartels, I., Seidel, J., Emmanuil, N., Petesen, M., Polityko, A., Dufke, A., Iourov, I., Trifonov, V., Vermeesch, J., Weise, A., 2007. Small supernumerary marker chromosomes (sSMC) in patients with a 45, X/46, X,+mar karyotype—17 new cases and a review of the literature. Sex Dev. 1, 353–362.

Liehr, T., Mrasek, K., Kosyakova, N., Ogilvie, C.M., Vermeesch, J., Trifonov, V., Rubtsov, N., 2008. Small supernumerary marker chromosomes (sSMC) in humans; are there B chromosomes hidden among them. Mol. Cytogenet. 1, 12.

Liehr, T., Weise, A., Hinreiner, S., Mkrtchyan, H., Mrasek, K., Kosyakova, N., 2010. Characterization of chromosomal rearrangements using multicolor-banding (MCB/m-band). Methods Mol. Biol. 659, 231–238.

Liehr, T., Bartels, I., Zoll, B., Ewers, E., Mrasek, K., Kosyakova, N., Merkas, M., Hamid, A.B., von Eggeling, F., Posorski, N., Weise, A., 2011. Is there a yet unreported unbalanced chromosomal abnormality without phenotypic consequences in proximal 4p? Cytogenet. Genome Res. 132, 121–123.

Liehr, T., Weise, A., Hamid, A.B., Fan, X., Klein, E., Aust, N., Othman, M.A.K., Mrasek, K., Kosyakova, N., 2013. Multicolor fluorescence in situ hybridization methods in nowadays clinical diagnostics. Expert. Rev. Mol. Diagn. 13, 251–255.

Lin, M.S., Zhang, A., Wilson, M.G., Fujimoto, A., 1988. DA/DAPI-fluorescent heteromorphism of human Y chromosome. Hum. Genet. 79, 36–38.

Lin, M.S., Huynh, K.H., Fujimoto, A., Wilson, M.G., 1990. Lack of specificity of DA/DAPI fluorescence. Clin. Genet. 37, 74–77.

Lin, M.S., Zhang, A., Fujimoto, A., Wilson, M.G., 1994. A rare 6q11+ heteromorphism: Cytogenetic analysis and in situ hybridization. Hum. Hered. 44, 31–36.

Lin, C.L., Gibson, L., Pober, B., Yang-Feng, T.L., 1995. A de novo satellited short arm of the Y chromosome possibly resulting from an unstable translocation. Hum. Genet. 96, 585–588.

Lindberg, L., Pelto, K., Borgström, G.H., 1992. Familial pericentric inversion (3)(p12q24). Hum. Genet. 89, 433–436.

Liu, S., Gao, C., Hu, Y., Liu, M., Cheng, Z., 1993. Molecular and clinical cytogenetic studies of a family with a 22p+ marker chromosome. Yi Chuan Xue Bao 20, 7–11.

Livingston, G.K., Lockey, J.E., Witt, K.S., Rogers, S.W., 1985. An unstable giant satellite associated with chromosomes 21 and 22 in the same individual. Am. J. Hum. Genet. 37, 553–560.

Lo, A.W., Liao, G.C., Rocchi, M., Choo, K.H., 1999. Extreme reduction of chromosome-specific alpha-satellite array is unusually common in human chromosome 21. Genome Res. 9, 895–908.

Looijenga, L.H.J., Oosterhuis, J.W., Smit, V.T.H.B.M., Wessels, J.W., Mollevanger, P., 1992. Devilee P Alpha satellite DNAs on chromosome 10 and 12 are both members of the dimeric suprachromosomal subfamily, but display little identity at the nucleotide sequence level. Genomics 13, 1125–1132.

López-Expósito, I., Bafalliu, J.A., Santos, M., Fuster, C., Puche-Mira, A., Guillén-Navarro, E., 2008. Intrachromosomal partial triplication of chromosome 13 secondary to a paternal duplication with mild phenotypic effect. Am. J. Med. Genet. A 146A, 1190–1194.

Lubs, H.A., Ruddle, F.H., 1970. Chromosomal abnormalities in the human population: estimation of rates based on New Haven newborn study. Science 169, 495–497.

Lubs, H.A., Ruddle, F.H., 1971. Chromosome polymorphism in American Negro and White populations. Nature 233, 134–136.

Lubs, H.A., Kinberling, W.J., Hecht, F., Patil, S.R., Brown, J., Gerald, P., Summitt, R.L., 1977. Racial differences in the frequency of Q and C chromosomal heteromorphisms. Nature 268, 631–633.

Lupski, J.R., Stankiewicz, P., 2005. Genomic disorders: Molecular mechanisms for rearrangements and conveyed phenotypes. PLoS Genet. 1, e49.

Lurie, I.W., Rumyantseva, N.V., Zaletajev, D.V., Gurevich, D.B., Korotkova, I.A., 1985. Trisomy 20p: Case report and genetic review. J. Genet. Hum. 33, 67–75.

MacDonald, I.M., Cox, D.M., 1985. Inversion of chromosome 2 (p11p13): Frequency and implications for genetic counselling. Hum. Genet. 69, 281–283.

Macera, M.J., Verma, R.S., Conte, R.A., Bialer, M.G., Klein, V.R., 1995. Mechanisms of the origin of a G-positive band within the secondary constriction region of human chromosome 9. Cytogenet. Cell Genet. 69, 235–239.

Madan, K., Bruinsma, A.H., 1979. C-band polymorphism in human chromosome no. 6. Clin. Genet. 15, 193–197.

Madon, P.F., Athalye, A.S., Parikh, F.R., 2005. Polymorphic variants on chromosomes probably play a significant role in infertility. Reprod. Biomed. Online 11, 726–732.

Madrigal, I., Martinez, M., Rodriguez-Revenga, L., Carrió, A., Milà, M., 2012. 12p13 rearrangements: 6 Mb deletion responsible for ID/MCA and reciprocal duplication without clinical responsibility. Am. J. Med. Genet. A 158A, 1071–1076.

Maegenis, R.E., Donlon, T.A., Wyandt, H.E., 1978. Giemsa-11 staining of chromosome 1: A newly described heteromorphism. Science 202, 64–65.

Makino, S., Muramoto, J.I., Tabata, S., 1966. A survey of a familial transmission of an anomalous autosome in group 13-15. Chromosoma. 18, 371–379.

Mandal, A.K., Prabhakara, K., Reddy, A.B., Devi, A.R., Panicker, S.G., 2003. Congenital glaucoma associated with 22p+ variant in a dysmorphic child. Indian J. Ophthalmol. 51, 355–357.

Manuelidis, L., 1978. Complex and simple sequences in human repeated DNAs. Chromosoma. 66, 1–21.

Manvelyan, M., Schreyer, I., Höls-Herpertz, I., Köhler, S., Niemann, R., Hehr, U., Belitz, B., Bartels, I., Götz, J., Huhle, D., Kossakiewicz, M., Tittelbach, H., Neubauer, S., Polityko, A., Mazauric, M.L., Wegner, R., Stumm, M., Küpferling, P., Süss, F., Kunze, H., Weise, A., Liehr, T., Mrasek, K., 2007. Forty-eight new cases with infertility due to balanced chromosomal rearrangements: Detailed molecular cytogenetic analysis of the 90 involved breakpoints. Int. J. Mol. Med. 19, 855–864.

Manvelyan, M., Riegel, M., Santos, M., Fuster, C., Pellestor, F., Mazaurik, M.L., Schulze, B., Polityko, A., Tittelbach, H., Reising-Ackermann, G., Belitz, B., Hehr, U., Kelbova, C., Volleth, M., Gödde, E., Anderson, J., Küpferling, P., Köhler, S., Duba, H.C., Dufke, A., Aktas, D., Martin, T., Schreyer, I., Ewers, E., Reich, D., Mrasek, K., Weise, A., Liehr, T., 2008. Thirty-two new cases with small supernumerary marker chromosomes detected in connection with fertility problems: Detailed molecular cytogenetic characterization and review of the literature. Int. J. Mol. Med. 21, 705–714.

Manvelyan, M., Cremer, F.W., Lancé, J., Kläs, R., Kelbova, C., Ramel, C., Reichenbach, H., Schmidt, C., Ewers, E., Kreskowski, K., Ziegler, M., Kosyakova, N., Liehr, T., 2011. New cytogenetically visible copy number variant in region 8q21.2. Mol. Cytogenet. 4, 1.

Manz, E., Alkan, M., Bühler, E., Schmidtke, J., 1992. Arrangement of DYZ1 and DYZ2 repeats on the human Y-chromosome: A case with presence of DYZ1 and absence of DYZ2. Mol. Cell Probes. 6, 257–259.

Manzanal Martínez, A.I., Robledo Batanero, M., Ramos Corrales, C., Ayuso García, C., Sánchez Cascos, A., 1992. Cytogenetic study of a family with 15p+ chromosomal plymorphism. An. Esp. Pediatr. 36, 269–271.

Maranda, B., Lemieux, N., Lemyre, E., 2006. Familial deletion 18p syndrome: Case report. BMC Med. Genet. 7, 60.

Marçais, B., Charlieu, J.P., Allain, B., Brun, E., Bellis, M., Roizès, G., 1991. On the mode of evolution of alpha satellite DNA in human populations. J. Mol. Evol. 33, 42–48.

Marçais, B., Laurent, A.M., Charlieu, J.P., Roizès, G., 1993. Organization of the variant domains of alpha satellite DNA on human chromosome 21. J. Mol. Evol. 37, 171–178.

Martin Lucas, M.A., Pérez Castillo, A., Abrisqueta, J.A., 1984. Origin and structure of a satellited Y chromosome. Ann. Genet. 27, 184–186.

Martínez, J.E., Tuck-Muller, C.M., Gasparrini, W., Li, S., Wertelecki, W., 1999. 1p microdeletion in sibs with minimal phenotypic manifestations. Am. J. Med. Genet. 82, 107–109.

Marzais, B., Vorsanova, S.G., Roizes, G., Yurov, Y.B., 1999. Analysis of alphoid DNA variation and kinetochore size in human chromosome 21: Evidence against pathological significance of alphoid satellite DNA diminutions. Tsitol. Genet. 33, 25–31.

Mashkova, T.D., Akopian, T.A., Romanova, L.Y., Mitkevich, S.P., Yurov, Y.B., Kisselev, L.L., Alexandrov, I.A., 1994. Genomic organization, sequence and polymorphism of the human chromosome 4-specific alpha-satellite DNA. Gene. 140, 211–217.

Mashkova, T., Oparina, N., Alexandrov, I., Zinovieva, O., Marusina, A., Yurov, Y., Lacroix, M.H., Kisselev, L., 1998. Unequal cross-over is involved in human alpha satellite DNA rearrangements on a border of the satellite domain. FEBS Lett. 441, 451–457.

Matera, A.G., Baldini, A., Ward, D.C., 1993. An oligonucleotide probe specific to the centromeric region of human chromosome 5. Gnomics. 18, 729–731.

Matsuda, T., Sanada, S., Omori, K., Horii, Y., Takahashi, Y., Edamura, S., Koike, S., Sasaki, M., 1986. Autosomal translocation and associated male infertility. Hinyokika Kiyo 32, 809–818.

McClintock, B., 1984. The significance of responses of the genome to challenge. Science 226, 792–801.

McDermid, H.E., Duncan, A.M., Higgins, M.J., Hamerton, J.L., Rector, E., Brasch, K.R., White, B.N., 1986. Isolation and characterization of an alpha-satellite repeated sequence from human chromosome 22. Chromosoma. 94, 228–234.

McKenzie, W.H., Lubs, H.A., 1974. Human Q and C chromosomal variants: Distribution and incidence. Cytogenet. Cell Genet. 14, 97–115.

Meneveri, R., Agresti, A., Marozzi, A., Saccone, S., Rocchi, M., Archidiacono, N., Corneo, G., Della Valle, G., Ginelli, E., 1993. Molecular organization and chromosomal location of human GC-rich heterochromatic blocks. Gene. 123, 227–234.

Metaxotou, C., Kalpini-Mavrou, A., Panagou, M., Tsenghi, C., 1978. Polymorphism of chromosome 9 in 600 Greek subjects. Am. J. Hum. Genet. 30, 85–89.

Metz, F., Bier, L., Pfeiffer, R.A., 1973. Partial trisomy of the short arm of chromosome 4 due to translocation t(4p-22p+). Humangenetik 18, 207–211.

Meza-Espinoza, J.P., Davalos-Rodríguez, I.P., Rivera-Ramírez, H., Perez-Muñoz, S., Rivas-Solís, F., 2006. Chromosomal abnormalities in patients with azoospermia in Western Mexico. Arch. Androl. 52, 87–90.

Mikelsaar, A.V., Tüür, S.J., Käosaar, M.E., 1973. Human karyotype polymorphism. I. Routine and fluorescence microscopic investigation of chromosomes in a normal adult population. Humangenetik 20, 89–101.

Mikelsaar, A.V., Käosaar, M.E., Tüür, S.J., Viikmaa, M.H., Talvik, T.A., Lääts, J., 1975. Human karyotype polymorphism. III. Routine ank fluorescence microscopic investigation of chromosomes in normal adults and mentally retarded children. Humangenetik 26, 1–23.

Miller, D.A., Tantravahi, R., Dev, V.G., Miller, O.J., 1977. Frequency of satellite association of human chromosomes is correlated with amount of Ag-staining of the nucleolus organizer region. Am. J. Hum. Genet. 29, 490–502.

Miller, D.A., Breg, W.R., Warburton, D., Dev, V.G., Miller, O.J., 1978. Regulation of rRNA gene expression in a human familial 14p+ marker chromosome. Hum. Genet. 43, 289–297.

Miller, I., Songster, G., Fontana, S., Hsieh, C.L., 1995. Satellited 4q identified in amniotic fluid cells. Am. J. Med. Genet. 55, 237–239.

Millington, K., Hudnall, S.D., Northup, J., Panova, N., Velagaleti, G., 2008. Role of chromosome 1 pericentric heterochromatin (1q) in pathogenesis of myelodysplastic syndromes: Report of 2 new cases. Exp. Mol. Pathol. 84, 189–193.

Mizunoe, T., Young, S.R., 1992. Low fluorescence alpha satellite region yields negative result. Prenat. Diagn. 12, 549–550.

Moog, U., Engelen, J.J., Albrechts, J.C., Baars, L.G., de Die-Smulders, C.E., 2000. Familial dup(8)(p12p21.1): Mild phenotypic effect and review of partial 8p duplications. Am. J. Med. Genet. 94, 306–310.

Moore, I.K., Martin, M.P., Dorsey, M.J., Paquin, C.E., 2000. Formation of circular amplifications in Saccharomyces cerevisiae by a breakage-fusion-bridge mechanism. Environ. Mol. Mutagen. 36, 113–120.

Morales, C., Soler, A., Bruguera, J., Madrigal, I., Alsius, M., Obon, M., Margarit, E., Sánchez, A., 2007. Pseudodicentric 22;Y translocation transmitted through four generations of a large family without phenotypic repercussion. Cytogenet. Genome Res. 116, 319–323.

Moreira, L.M., Riegel, M., 2000. Two sibs with duplication of 4q31–>qter due to 3:1 meiotic disjunction and mild phenotype. Genet. Couns. 11, 249–259.

Morerio, C., Rapella, A., Tassano, E., Lanino, E., Micalizzi, C., Rosanda, C., Panarello, C., 2006. Gain of 1q in pediatric myelodysplastic syndromes. Leuk. Res. 30, 1437–1441.

Morris, M.I., Hanson, F.W., Tennant, F.R., 1987. A novel Y/13 familial translocation. Am. J. Obstet. Gynecol. 157, 857–858.

Mrasek, K., Krüger, G., Bauer, I., Müller-Navia, J., Liehr, T., Weise, A., 2008. A new unbalanced chromosomal abnormality in 1q31.1 to 1q32 without phenotypic consequences. Cytogenet. Genome Res. 121, 286–287.

Mrasek, K., Schoder, C., Teichmann, A.C., Behr, K., Franze, B., Wilhelm, K., Blaurock, N., Claussen, U., Liehr, T., Weise, A., 2010. Global screening and extended nomenclature for 230 aphidicolin-inducible fragile sites, including 61 yet unreported ones. Int. J. Oncol. 36, 929–940.

Müllenbach, R., Pusch, C., Holzmann, K., Suijkerbuijk, R., Blin, N., 1996. Distribution and linkage of repetitive clusters from the heterochromatic region of human chromosome 22. Chromosome Res. 4, 282–287.

Murthy, D.S., Murthy, S.K., Patel, J.K., Tyagi, A.A., 1989. Nucleolar organizer region heteromorphism associated with trisomy-21: A risk factor for non-disjunction? Indian J. Exp. Biol. 27, 864–867.

Murthy, D.S., Sundareshan, T.S., Farag, T.I., al-Awadi, S.A., al-Othman, S.A., 1990. Segregation of acrocentric chromosome association in familial dicentric Robertsonian translocation t(14p;22p), aneuploidy (trisomy-21) and heteromorphism. Indian J. Exp. Biol. 28, 511–515.

Musilova, P., Rybar, R., Oracova, E., Veselá, K., Rubes, J., 2008. Hybridization of the 18 alpha-satellite probe to chromosome 1 revealed in PGD. Reprod. Biomed. Online 17, 695–698.

Mutton, D.E., Daker, M.G., 1973. Pericentric inversion of chromosome 9. Nat. New Biol. 241, 80.

Myokai, F., Takashiba, S., Lebo, R., Amar, S., 1999. A novel lipopolysaccharide-induced transcription factor regulating tumor necrosis factor alpha gene expression: molecular cloning, sequencing, characterization, and chromosomal assignment. Proc. Natl. Acad. Sci. U. S. A. 96, 4518–4523.

Neglia, M., Bertoni, L., Zoli, W., Giulotto, E., 2003. Amplification of the pericentromeric region of chromosome 1 in a newly established colon carcinoma cell line. Cancer Genet. Cytogenet. 142, 99–106.

Neumann, A.A., Robson, L.G., Smith, A.A., 1992. 15p+ variant shown to be a t(Y;15) with fluorescence in situ hybridisation. Ann. Genet. 35, 227–230.

Ng, L.K., Kwok, Y.K., Tang, L.Y., Ng, P.P., Ghosh, A., Lau, E.T., Tang, M.H., 2006. Prenatal detection of a de novo Yqh-acrocentric translocation. Clin. Biochem. 39, 219–223.

Nielsen, J., Friedrich, U., Hreidarsson, A.B., 1974. Frequency and genetic effect of 1qh plus. Humangenetik 21, 193–196.

Nielsen, J., Friedrich, U., Hreidarsson, A.B., Zeuthen, E., 1974. Frequency of 9qh+ and risk of chromosome aberrations in the progeny of individuals with 9qh+. Humangenetik 21, 211–216.

Nielsen, J., Sillesen, I., 1975. Incidence of chromosome aberrations among 11148 newborn children. Humangenetik 30, 1–12.

Nielsen, J., Hreidarsson, A.B., Brynjolfsson, T., Holm, V., Madsen, A.M., Petersen, G.B., Lamm, L., 1978. Saldaña-Garcia P. A family with a chromosome 14 short arm deletion. Hereditas 88, 107–112.

Nilsson, M., Krejci, K., Koch, J., Kwiatkowski, M., Gustavsson, P., Landegren, U., 1997. Padlock probes reveal single-nucleotide differences, parent of origin and in situ distribution of centromeric sequences in human chromosomes 13 and 21. Nat. Genet. 16, 252–255.

Norris, F.M., Mercer, B., Pertile, M.D., 1995. Interstitial insertion of NORs into Yq and 22q: Two case studies. Bull. Hum. Genet. Soc. Australas 8, 48.

Nowell, P.C., Hungerford, D.A., 1960. A minute chromosome in human chronic granulocytic leukemia. Science 132, 1497.

Nusbaum, C., Mikkelsen, T.S., Zody, M.C., Asakawa, S., Taudien, S., Garber, M., Kodira, C.D., Schueler, M.G., Shimizu, A., Whittaker, C.A., Chang, J.L., Cuomo, C.A., Dewar, K., FitzGerald, M.G., Yang, X., Allen, N.R., Anderson, S., Asakawa, T., Blechschmidt, K., Bloom, T., Borowsky, M.L., Butler, J., Cook, A., Corum, B., DeArellano, K., DeCaprio, D., Dooley, K.T., Dorris 3rd, L., Engels, R., Glöckner, G., Hafez, N., Hagopian, D.S., Hall, J.L., Ishikawa, S.K., Jaffe, D.B., Kamat, A., Kudoh, J., Lehmann, R., Lokitsang, T., Macdonald, P., Major, J.E., Matthews, C.D., Mauceli, E., Menzel, U., Mihalev, A.H., Minoshima, S., Murayama, Y., Naylor, J.W., Nicol, R., Nguyen, C., O'Leary, S.B., O'Neill, K., Parker, S.C., Polley, A., Raymond, C.K., Reichwald, K., Rodriguez, J., Sasaki, T., Schilhabel, M., Siddiqui, R., Smith, C.L., Sneddon, T.P., Talamas, J.A., Tenzin, P., Topham, K., Venkataraman, V., Wen, G., Yamazaki, S., Young, S.K., Zeng, Q., Zimmer, A.R., Rosenthal, A., Birren, B.W., Platzer, M., Shimizu, N., Lander, E.S., 2006. DNA sequence and analysis of human chromosome 8. Nature 439, 331–335.

O'Keefe, C.L., Matera, A.G., 2000. Alpha satellite DNA variant-specific oligoprobes differing by a single base can distinguish chromosome 15 homologs. Genome Res. 10, 1342–1350.

O'Keefe, C.L., Griffin, D.K., Bean, C.J., Matera, A.G., Hassold, T.J., 1997. Alphoid variant-specific FISH probes can distinguish autosomal meiosis I from meiosis II nondisjunction in human sperm. Hum. Genet. 101, 61–66.

Okamoto, E., Miller, D.A., Erlanger, B.F., Miller, O.J., 1981. Polymorphism of 5-methylcytosine-rich DNA in human acrocentric chromosomes. Hum. Genet. 58, 255–259.

Onrat, S.T., Söylemez, Z., Elmas, M., 2012. 46, XX, der(15), t(Y;15)(q12;p11) karyotype in an azoospermic male. Indian J. Hum. Genet. 18, 241–245.

Otter, M., Schrander-Stumpel, C.T., Didden, R., Curfs, L.M., 2012. The psychiatric phenotype in triple X syndrome: New hypotheses illustrated in two cases. Dev. Neurorehabil. 15, 233–238.

Ozdemir, M., Yuksel, Z., Karaer, K., Tekin, N., Kucuk, H., Erzurumluoglu, E., Cilingir, O., 2012. Partial trisomies of 8q and 15q due to maternal balanced translocations. Genet. Couns. 23, 375–382.

Paar, V., Basar, I., Rosandić, M., Gluncić, M., 2007. Consensus higher order repeats and frequency of string distributions in human genome. Curr. Genomics 8, 93–111.

Parcheta, B., Skawiński, W., Wiśniewski, L., Piontek, E., Gutkowska, A., Wermeński, K., 1985. A new case of partial trisomy of 17 long arm. Densitometric analysis of aberrations. Eur. J. Pediatr. 143, 314–316.

Park, J.P., Rawnsley, B.E., 1996. Prenatal detection of chromosome 20 variants (20ph+, 20ps). Prenat. Diagn. 16, 771–774.

Park, V.M., Gustashaw, K.M., Wathen, T.M., 1992. The presence of interstitial telomeric sequences in constitutional chromosome abnormalities. Am. J. Hum. Genet. 50, 914–923.

Pathak, S., 1979. Cytogenetic research techniques in humans and laboratory animals that can be applied most profitably to livestock. J. Dairy Sci. 62, 836–843.

Patil, S.R., Bent, F.C., 1980. Silver staining and the 17ps chromosome. Clin. Genet. 17, 281–284.

Pellestor, F., Girardet, A., Andréo, B., Charlieu, J.P., 1994. A polymorphic alpha satellite sequence specific for human chromosome 13 detected by oligonucleotide primed in situ labelling (PRINS). Hum. Genet. 94, 346–348.

Penna Videaú, S., Araujo, H., Ballesta, F., Ballescá, J.L., Vanrell, J.A., 2001. Chromosomal abnormalities and polymorphisms in infertile men. Arch. Androl. 46, 205–210.

Percy, M.E., Dearie, T.G., Jabs, E.W., Bauer, S.J., Chodakowski, B., Somerville, M.J., Lennox, A., McLachlan, D.R., Baldini, A., Miller, D.A., 1993. Family with 22-derived marker chromosome and late-onset dementia of the Alzheimer type: II. Further cytogenetic analysis of the marker and characterization of the high-level repeat sequences using fluorescence in situ hybridization. Am. J. Med. Genet. 47, 14–19.

Pérez-Castillo, A., Martín-Lucas, M.A., Abrisqueta, J.A., 1986. New insights into the effects of extra nucleolus organizer regions. Hum. Genet. 72, 80–82.

Petersen, M.B., 1986. Rare chromosome 20 variants encountered during prenatal diagnosis. Prenat. Diagn. 6, 363–367.

Peterson, M.B., Frantzen, M., Antonarakis, S.E., Warren, A.C., Van Broeckhoven, C., Chakravarti, A., Cox, T.K., Lund, C., Olsen, B., Poulsen, H., Sand, A., Tommerup, N., Mikkelsen, M., 1992. Comparative study of microsatellite and cytogenetic markers for detecting the origin of the nondisjoined chromosome 21 in Down syndrome. Am. J. Hum. Genet. 51, 516–525.

Petit, P., Devriendt, K., Vermeesch, J.R., Meireleire, J., Fryns, J.P., 1998. Localization by FISH of centric fission breakpoints in a de novo trisomy 9p patient with i(9p) and t(9q;11p). Genet. Couns. 9, 215–221.

Piard, J., Philippe, C., Marvier, M., Beneteau, C., Roth, V., Valduga, M., Béri, M., Bonnet, C., Grégoire, M.J., Jonveaux, P., Leheup, B., 2010. Clinical and molecular characterization of a large family with an interstitial 15q11q13 duplication. Am. J. Med. Genet. A 152A, 1933–1941.

Piccini, I., Ballarati, L., Bassi, C., Rocchi, M., Marozzi, A., Ginelli, E., Meneveri, R., 2001. The structure of duplications on human acrocentric chromosome short arms derived by the analysis of 15p. Hum. Genet. 108, 467–477.

Pinkel, D., Straume, T., Gray, J.W., 1986. Cytogenetic analysis using quantitative, high-sensitivity, fluorescence hybridization. Proc. Natl. Acad. Sci. USA 83, 2934–2938.

Pimentel, D., Alonso, P., Abrisqueta, J.A., 1989. Unusual chromosome 20 anomaly arising "de novo" to give dic(20)qs. Hum. Genet. 84, 97–98.

Pittalis, M.C., Santarini, L., Bovicelli, L., 1994. Prenatal diagnosis of a heterochromatic 18p+ heteromorphism. Prenat. Diagn. 14, 72–73.

Polityko, A.D., Khurs, O.M., Kulpanovich, A.I., Mosse, K.A., Solntsava, A.V., Rumyantseva, N.V., Naumchik, I.V., Liehr, T., Weise, A., Mkrtchyan, H., 2009. Paternally derived der(7)t(Y;7) (p11.1~11.2;p22.3)dn in a mosaic case with Turner syndrome. Eur. J. Med. Genet. 52, 207–210.

Pontier, D.B., Gribnau, J., 2011. Xist regulation and function explored. Hum. Genet. 130, 223–236.

Prades, C., Laurent, A.M., Puechberty, J., Yurov, Y., Roizés, G., 1996. SINE and LINE within human centromeres. J. Mol. Evol. 42, 37–43.

Prieto, F., Badía, L., Beneyto, M., Palau, F., 1989. Nucleolus organizer regions (NORs) inserted in 6q15. Hum. Genet. 81, 289–290.

Puechberty, J., Laurent, A.M., Gimenez, S., Billault, A., Brun-Laurent, M.E., Calenda, A., Marçais, B., Prades, C., Ioannou, P., Yurov, Y., Roizès, G., 1999. Genetic and physical analyses of the centromeric and pericentromeric regions of human chromosome 5: Recombination across 5cen. Genomics 56, 274–287.

Quack, B., Noël, B., Moine, A., 1987. Chromosome 18 variant with increased centromeric tandemly repeated DNA in a family. Ann. Genet. 30, 85–90.

Queralt, R., Madrigal, I., Vallecillos, M.A., Morales, C., Ballescá, J.L., Oliva, R., Soler, A., Sánchez, A., Margarit, E., 2008. Atypical XX male with the SRY gene located at the long arm of chromosome 1 and a 1qter microdeletion. Am. J. Med. Genet. A 146A, 1335–1340.

Quiroga, R., Monfort, S., Oltra, S., Ferrer-Bolufer, I., Roselló, M., Mayo, S., Martinez, F., Orellana, C., 2011. Partial duplication of 18q including a distal critical region for Edwards Syndrome in a patient with normal phenotype and oligoasthenospermia: Case report. Cytogenet. Genome Res. 133, 78–83.

Rajcan-Separovic, E., Robinson, W.P., Stephenson, M., Pantzar, T., Arbour, L., McFadden, D., Guscott, J., 2001. Recurrent trisomy 15 in a female carrier of der(15) t(Y;15)(q12;p13). Am. J. Med. Genet. 99, 320–324.

Rahman, M.M., Bashamboo, A., Prasad, A., Pathak, D., Ali, S., 2004. Organizational variation of DYZ1 repeat sequences on the human Y chromosome and its diagnostic potentials. DNA Cell Biol. 23, 561–571.

Ramos, S., Alcántara, M.A., Molina, B., del Castillo, V., Sánchez, S., Frias, S., 2008. Acrocentric cryptic translocation associated with nondisjunction of chromosome 21. Am. J. Med. Genet. A 146A, 97–102.

Reddy, K.S., Sulcova, V., 1998. The mobile nature of acrocentric elements illustrated by three unusual chromosome variants. Hum. Genet. 102, 653–662.

Redon, R., Ishikawa, S., Fitch, K.R., Feuk, L., Perry, G.H., Andrews, T.D., Fiegler, H., Shapero, M.H., Carson, A.R., Chen, W., Cho, E.K., Dallaire, S., Freeman, J.L., González, J.R., Gratacòs, M., Huang, J., Kalaitzopoulos, D., Komura, D., MacDonald, J.R., Marshall, C.R., Mei, R., Montgomery, L., Nishimura, K., Okamura, K., Shen, F., Somerville, M.J., Tchinda, J., Valsesia, A., Woodwark, C., Yang, F., Zhang, J., Zerjal, T., Zhang, J., Armengol, L., Conrad, D.F., Estivill, X., Tyler-Smith, C., Carter, N.P., Aburatani, H., Lee, C., Jones, K.W., Scherer, S.W., Hurles, M.E., 2006. Global variation in copy number in the human genome. Nature 444, 444–454.

Ren, H., Francis, W., Boys, A., Chueh, A.C., Wong, N., La, P., Wong, L.H., Ryan, J., Slater, H.R., Choo, K.H., 2005. BAC-based PCR fragment microarray: High-resolution detection of chromosomal deletion and duplication breakpoints. Human Mutation 25, 476–482.

Riordan, D., Dawson, A.J., 1998. The evaluation of 15q proximal duplications by FISH. Clin. Genet. 54, 517–521.

Rizzi, N., Denegri, M., Chiodi, I., Corioni, M., Valgardsdottir, R., Cobianchi, F., Riva, S., Biamonti, G., 2004. Transcriptional activation of a constitutive heterochromatic domain of the human genome in response to heat shock. Mol. Biol. Cell 15, 543–551.

Roa, B.B., Lupski, J.R., 1994. Molecular genetics of Charcot-Marie-Tooth neuropathy. Adv. Hum. Genet. 22, 117–152.

Rocchi, M., Baldini, A., Archidiacono, N., Lainwala, S., Miller, O.J., Miller, D.A., 1990. Chromosome-specific subsets of human alphoid DNA identified by a chromosome 2-derived clone. Genomics 8, 705–709.

Rocchi, M., Archidiacono, N., Ward, D.C., Baldini, A., 1991. A human chromosome 9-specific alphoid DNA repeat spatially resolvable from satellite 3 DNA by fluorescent in situ hybridization. Genomics 9, 517–523.

Rodríguez, L., Zollino, M., Mansilla, E., Martínez-Fernández, M.L., Pérez, P., Murdolo, M., Martínez-Frías, M.L., 2007. The first 4p euchromatic variant in a healthy carrier having an unusual reproductive history. Am. J. Med. Genet. A 143A, 995–998.

Romain, D.R., Whyte, S., Callen, D.F., Eyre, H.J., 1991. A rare heteromorphism of chromosome 20 and reproductive loss. J. Med. Genet. 28, 477–478.

Roos, A., Elbracht, M., Baudis, M., Senderek, J., Schönherr, N., Eggermann, T., Schüler, H.M., 2008. A 10.7 Mb interstitial deletion of 13q21 without phenotypic effect defines a further non-pathogenic euchromatic variant. Am. J. Med. Genet. A 146A, 2417–2420.

Rosenfeld, J.A., Traylor, R.N., Schaefer, G.B., McPherson, E.W., Ballif, B.C., Klopocki, E., Mundlos, S., Shaffer, L.G., Aylsworth, A.S., 2012. 1q21.1 Study Group. Proximal microdeletions and microduplications of 1q21.1 contribute to variable abnormal phenotypes. Eur. J. Hum. Genet. 20, 754–761.

Rossi, E., de Gregori, M., Grazia Patricelli, M., Pramparo, T., Argentiero, L., Giglio, S., Sosta, K., Foresti G, Zuffardi, O., 2005. 8.5 Mb deletion at distal 5p in a male ascertained for azoospermia. Am. J. Med. Genet. A 133A, 189–192.

Roux, C., Taillemite, J.L., Baheux-Morlier, G., 1974. Partial trisomy 10q dueto familial translocation t(10q-; 22p-plus). Ann. Genet. 17, 59–62.

Rudd, M.K., Wray, G.A., Willard, H.F., 2006. The evolutionary dynamics of alpha-satellite. Genome Res. 16, 88–96.

Sala, E., Villa, N., Crosti, F., Miozzo, M., Perego, P., Cappellini, A., Bonazzi, C., Barisani, D., Dalprà, L., 2007. Endometrioid-like yolk sac and Sertoli-Leydig cell tumors in a carrier of a Y heterochromatin insertion into 1qh region: A causal association? Cancer Genet. Cytogenet. 173, 164–169.

Sánchez, L., Martínez, P., Goyanes, V., 1991. Analysis of centromere size in human chromosomes 1, 9, 15, and 16 by electron microscopy. Genome 34, 710–713.

Sands, V.E., 1969. Short arm enlargement in acrocentric chromosomes. Am. J. Hum. Genet. 21, 293–304.

Sanger, T.M., Olney, A.H., Zaleski, D., Pickering, D., Nelson, M., Sanger, W.G., Dave, B.J., 2005. Cryptic duplication and deletion of 9q34.3 –> qter in a family with a t(9;22)(q34.3;p11.2). Am. J. Med. Genet. A 138, 51–55.

Sarri, C., Douzgou, S., Gyftodimou, Y., Tümer, Z., Ravn, K., Pasparaki, A., Sarafidou, T., Kontos, H., Kokotas, H., Karadima, G., Grigoriadou, M., Pandelia, E., Theodorou, V., Moschonas, N.K., Petersen, M.B., 2011. Complex distal 10q rearrangement in a girl with mild intellectual disability: Follow up of the patient and review of the literature of non-acrocentric satellited chromosomes. Am. J. Med. Genet. A 155A, 2841–2854.

Savary, J.B., Vasseur, F., Manouvrier, S., Daudignon, A., Lemaire, O., Thieuleux, M., Poher, M., Lequien, P., Deminatti, M.M., 1991. Trisomy 16q23—qter arising from a maternal t(13;16)(p12;q23): case report and evidence of the reciprocal balanced maternal rearrangement by the Ag-NOR technique. Hum. Genet. 88, 115–118.

Savelyeva, L., Schneider, B., Finke, L., Schlag, P., Schwab, M., 1994. Amplification of satellite DNA at 16q11.2 in the germ-line of a patient with breast-cancer. Int. J. Oncol. 4, 347–351.

Schempp, W., Weber, B., Serra, A., Neri, G., Gal, A., Wolf, U.A., 1985. 45, X male with evidence of a translocation of Y euchromatin onto chromosome 15. Hum. Genet. 71, 150–154.

Schinzel, A., 1982. The use of chromosome variants in clinical cytogenetics. Wien Klin Wochenschr 94, 210–213.

Schinzel, A., 2001. Catalogue of unbalanced chromosome aberrations in man, second ed. De Gruyter, Berlin, New York.

Schlegelberger, B., 2013. In memoriam: Prof. Dr. rer. nat. Dr. med. h.c. Lore Zech; 24.9.1923 - 13.3.2013: Honorary member of the European Society of Human Genetics, Honorary member of the German Society of Human Genetics, Doctor laureate, the University of Kiel, Germany. Mol. Cytogenet. 6, 20.

Schmid, M., Haaf, T., Solleder, E., Schempp, W., Leipoldt, M., Heilbronner, H., 1984. Satellited Y chromosomes: structure, origin, and clinical significance. Hum. Genet. 67, 72–85.

Schmid, M., Guttenbach, M., Nanda, I., Studer, R., Epplen, J.T., 1990. Organization of DYZ2 repetitive DNA on the human Y chromosome. Genomics 6, 212–218.

Schmid, M., Nanda, I., Steinlein, C., Epplen, J.T., 1994. Amplification of (GACA)n simple repeats in an exceptional 14p+ marker chromosome. Hum. Genet. 93, 375–382.

Schoumans, J., Sanner, G., Nordenskjöld, M., Anderlid, B.M., 2005. Detailed clinical description of four patients with 1.3 and 2.1 Mb chromosome imbalances derived from a familial t(12;17)(q24.33;q25.3). Am. J. Med. Genet. A 134, 254–258.

Schueler, M.G., Higgins, A.W., Rudd, M.K., Gustashaw, K., Willard, H.F., 2001. Genomic and genetic definition of a functional human centromere. Science 294, 109–115.

Schwanitz, G., 1976. Die Normvarianten menschlicher Chromosomen. Verlag Dr. med. D. Straube, Erlangen.

Schwanitz, G., Grosse, K.P., 1973. Partial trisomy 4p with translocation 4p-, 22p+ in the father. Ann. Genet. 16, 263–266.

Schwanitz, G., Schmid, R.D., Grosse, G., Grahn-Liebe, E., 1977. Familial translocation 3/22 MAT with partial trisomy 3q. J. Genet. Hum. 25, 141–150.

Schlattl, A., Anders, S., Waszak, S.M., Huber, W., Korbel, J.O., 2011. Relating CNVs to transcriptome data at fine resolution: Assessment of the effect of variant size, type, and overlap with functional regions. Genome Res. 21, 2004–2013.

Sensi, A., Giunta, C., Bonfatti, A., Gruppioni, R., Rubini, M., Fontana, F., 1994. Heteromorphic variant 18ph+ analyzed by sequential CBG and fluorescence in situ hybridization. Hum. Hered. 44, 295–297.

Serakinci, N., Pedersen, B., Koch, J., 2001. Expansion of repetitive DNA into cytogenetically visible elements. Cytogenet. Cell Genet. 92, 182–185.

Shabtai, F., Eilam, N., Elian, E., Halbrecht, I., 1981. A new family with a satellited Y. Ann. Genet. 24, 223–225.

Shah, H.A., Verma, R.S., Conte, R.A., Chester, M., Shklovskaya, T.V., Kleymann, S.M., Diaz-Barrios, V., Feldman, B., Lin, J.H., Serman, J., 1997. Fishing for the origin of satellite on the long arm of chromosome 4. Am. J. Hum. Genet. 61 (Suppl.), A375.

She, X., Horvath, J.E., Jiang, Z., Liu, G., Furey, T.S., Christ, L., Clark, R., Graves, T., Gulden, C.L., Alkan, C., Bailey, J.A., Sahinalp, C., Rocchi, M., Haussler, D., Wilson, R.K., Miller, W., Schwartz, S., Eichler, E.E., 2004. The structure and evolution of centromeric transition regions within the human genome. Nature 430, 857–864.

Sheth, F., Gohel, N., Liehr, T., Akinde, O., Desai, M., Adeteye, O., Sheth, J., 2012. Gain of chromosome 4qter and loss of 5pter – An unusual case with features of cri du chat syndrome (CdCS). Case Rep. Genet. 2012, 153405.

Shiels, C., Coutelle, C., Huxley, C., 1997. Contiguous arrays of satellites 1, 3, and beta form a 1.5-Mb domain on chromosome 22p. Genomics 44, 35–44.

Shim, S.H., Pan, A., Huang, X.L., Tonk, V.S., Varma, S.K., Milunsky, J.M., Wyandt, H.E., 2003. FISH Variants with D15Z1. J. Assoc. Genet. Technol. 29, 146–151.

Shinawi, M., Cheung, S.W., 2008. The array CGH and its clinical applications. Drug Discov. Today 13, 760–770.

Shumaker, D.K., Dechat, T., Kohlmaier, A., Adam, S.A., Bozovsky, M.R., Erdos, M.R., Eriksson, M., Goldman, A.E., Khuon, S., Collins, F.S., Jenuwein, T., Goldman, R.D., 2006. Mutant nuclear lamin A leads to progressive alterations of epigenetic control in premature aging. Proc. Natl. Acad. Sci. USA 103, 8703–8708.

Simi, P., Voliani, S., Rossi, S., Olivieri, L., Menchini Fabris, F., 1991. Pericentric inversion of chromosome 5: A possible threat to male fertility? Acta. Eur. Fertil. 22, 117–119.

Sipos, B., Massingham, T., Stütz, A.M., Goldman, N., 2012. An improved protocol for sequencing of repetitive genomic regions and structural variations using mutagenesis and next generation sequencing. PLoS One 7, e43359.

Sirvent, N., Forus, A., Lescaut, W., Burel, F., Benzaken, S., Chazal, M., Bourgeon, A., Vermeesch, J.R., Myklebost, O., Turc-Carel, C., Ayraud, N., Coindre, J.M., Pedeutour, F., 2000. Characterization of centromere alterations in liposarcomas. Genes Chromosomes Cancer 29, 117–129.

Skinner, J.I., Govberg, I.J., DePalma, R.T., Cotter, P.D., 2001. Heteromorphisms of chromosome 18 can obscure detection of fetal aneuploidy by interphase FISH. Prenat. Diagn. 21, 702–703.

Smeets, D.F., Merkx, G.F., Hopman, A.H., 1991. Frequent occurrence of translocations of the short arm of chromosome 15 to other D-group chromosomes. Hum. Genet. 87, 45–48.

Smeets, D., van Ravenswaaij, C., de Pater, J., Gerssen-Schoorl, K., Van Hemel, J., Janssen, G., Smits, A., 1997. At least nine cases of trisomy 11q23->qter in one generation as a result of familial t(11;13) translocation. J. Med. Genet. 34, 18–23.

Smith, A., Fraser, I.S., Elliot, G., 1979. An infantile male with balanced Y;19 translocation: Review of Y; autosome translocations. Ann. Genet. 22, 189–194.

Sneddon, T.P., Church, D.M., 2012. Online resources for genomic structural variation. Methods Mol. Biol. 838, 273–289.

Sokolic, R.A., Ferguson, W., Mark, H.F., 1999. Discordant detection of monosomy 7 by GTG-banding and FISH in a patient with Shwachman-Diamond syndrome without evidence of myelodysplastic syndrome or acute myelogenous leukemia. Cancer Genet. Cytogenet. 115, 106–113.

Soloviev, I.V., Yurov Yu, D., Vorsanova, S.G., Malet, P., Zerova, T.E., Buzhievskaya, T.I., 1998. Double color in situ hybridization of alpha-satellite chromosome 13, 21 specific cosmid clones for a rapid screening of their specificity. Tsitol. Genet. 32, 60–64.

Soudek, D., McCreary, B.D., Laraya, P., Dill, F.J., 1973 Jun. Two kinships with accessory bisatellited chromosomes. Ann. Genet. 16 (2), 101–107.

Soudek, D., Sroka, H., 1978. Inversion of 'flourescent' segment in chromosome 3: A polymorphic trait. Hum. Genet. 44, 109–115.

Soudek, D., 1979. Prenatal diagnosis of a 13p+ karyotype. Hum. Genet. 51, 339–341.

Spak, D.K., Johnston, K., Donlon, T.A., 1989. A non-centromeric C band variant on chromosome 11q23.2. J. Med. Genet. 26, 532–534.

Spinner, N.B., Biegel, J.A., Sovinsky, L., McDonald-McGinn, D., Rehberg, K., Parmiter, A.H., Zackai, E.H., 1993. 46, XX,15p+ documented as dup (17p) by fluorescence in situ hybridization. Am. J. Med. Genet. 46, 95–97.

Spowart, G., 1978. Triplication of a short arm region of chromosome 15. J. Med. Genet. 15, 404–405.

Spowart, G., 1979. Reassessment of presumed Y/22 and Y/15 translocations in man using a new technique. Cytogenet. Cell Genet. 23, 90–94.

Spreiz, A., Müller, D., Zotter, S., Albrecht, U., Baumann, M., Fauth, C., Erdel, M., Zschocke, J., Utermann, G., Kotzot, D., 2010. Phenotypic variability of a deletion and duplication 6q16.1->q21 due to a paternal balanced ins(7;6)(p15;q16.1q21). Am. J. Med. Genet. A 152A, 2762–2767.

Spurdle, A., Jenkins, T., 1992. The inverted Y-chromosome polymorphism in the Gujarati Muslim Indian population of South Africa has a single origin. Hum. Hered. 42, 330–332.

Srebniak, M.I., Boter, M., Verboven-Peerden, C.M., Looye-Bruinsma, G.A., Oudesluijs, G., Galjaard, R.J., Van Opstal, D., 2011. Prenatally diagnosed submicroscopic familial aberrations at 18p11.32 without phenotypic effect. Mol. Cytogenet. 4, 27.

Stankiewicz, P., Kuechler, A., Eller, C.D., Sahoo, T., Baldermann, C., Lieser, U., Hesse, M., Gläser, C., Hagemann, M., Yatsenko, S.A., Liehr, T., Horsthemke, B., Claussen, U., Marahrens, Y., Lupski, J.R., Hansmann, I., 2006. Minimal phenotype in a girl with trisomy 15q due to t(X;15)(q22.3;q11.2) translocation. Am. J. Med. Genet. A 140, 442–452.

Starke, H., Seidel, J., Henn, W., Reichardt, S., Volleth, M., Stumm, M., Behrend, C., Sandig, K.R., Kelbova, C., Senger, G., Albrecht, B., Hansmann, I., Heller, A., Claussen, U., Liehr, T., 2002. Homologous sequences at human chromosome 9 bands p12 and q13-21.1 are involved in different patterns of pericentric rearrangements. Eur. J. Hum. Genet. 10, 790–800.

Starke, H., Mrasek, K., Liehr, T., 2005. Three cases with enlarged acrocentric p-arms and two cases with cryptic partial trisomies. J. Histochem. Cytochem. 53, 359–360.

Steffensen, T.S., Gilbert-Barness, E., Sandstrom, M., Bell, J.R., Bryan, J., Sutcliffe, M.J., 2009. Extreme variant of enlarged heterochromatin region on chromosome 9Q in a normal child and multiple family members. Fetal. Pediatr. Pathol. 28, 247–252.

Stergianou, K., Gould, C.P., Waters, J.J., Hultén, M.A., 1993. A DA/DAPI positive human 14p heteromorphism defined by fluorescence in-situ hybridisation using chromosome 15-specific probes D15Z1 (satellite III) and p-TRA-25 (alphoid). Hereditas 119, 105–110.

Stetten, G., Sroka, B., Schmidt, M., Axelman, J., Migeon, B.R., 1986. Translocation of the nucleolus organizer region to the human X chromosome. Am. J. Hum. Genet. 39, 245–252.

Storto, P.D., Diehn, T.N., O'Malley, D.P., Bullard, B.A., Netzloff, M.L., VanDyke, D.L., Feldman, G.L., Precht, K.S., Ledbetter, D.H., Lese, C.M., 1999. Satellited chromosome 10 detected prenatally in a fetus and confirmed as mosaic in a parent. Prenat. Diagn. 19, 1088–1089.

Stochholm, K., Juul, S., Gravholt, C.H., 2012. Socio-economic factors affect mortality in 47, XYY syndrome-A comparison with the background population and Klinefelter syndrome. Am. J. Med. Genet. A 158A, 2421–2429.

Storto, P.D., Diehn, T.N., O'Malley, D.P., Bullard, B.A., Netzloff, M.L., VanDyke, D.L., Feldman, G.L., Precht, K.S., Ledbetter, D.H., Lese, C.M., 1999. Satellited chromosome 10 detected prenatally in a fetus and confirmed as mosaic in a parent. Prenat. Diagn. 19, 1088–1089.

Sullivan, B.A., Schwartz, S., Willard, H.F., 1996. Centromeres of human chromosomes. Environ. Mol. Mutagen. 28, 182–191.

Sumner, A.T., Evans, H.J., Buckland, R.A., 1971. New technique for distinguishing between human chromosomes. Nat. New Biol. 232, 31–32.

Sutton, V.R., Coveler, K.J., Lalani, S.R., Kashork, C.D., Shaffer, L.G., 2002. Subtelomeric FISH uncovers trisomy 14q32: lessons for imprinted regions, cryptic rearrangements and variant acrocentric short arms. Am. J. Med. Genet. 112, 23–27.

Swaminathan, G.J., Bragin, E., Chatzimichali, E.A., Corpas, M., Bevan, A.P., Wright, C.F., Carter, N.P., Hurles, M.E., Firth, H.V., 2012. DECIPHER: Web-based, community resource for clinical interpretation of rare variants in developmental disorders. Hum. Mol. Genet. 21, R37–R44.

Tagarro, I., Wiegant, J., Raap, A.K., González-Aguilera, J.J., Fernández-Peralta, A.M., 1994. Assignment of human satellite 1 DNA as revealed by fluorescent in situ hybridization with oligonucleotides. Hum. Genet. 93, 125–128.

Tamagaki, A., Shima, M., Tomita, R., Okumura, M., Shibata, M., Morichika, S., Kurahashi, H., Giddings, J.C., Yoshioka, A., Yokobayashi, Y., 2000. Segregation of a pure form of spastic paraplegia and NOR insertion into Xq11.2. Am. J. Med. Genet. 94, 5–8.

Tantravahi, U., Breg, W.R., Wertelecki, V., Erlanger, B.F., Miller, O.J., 1981. Evidence for methylation of inactive human rRNA genes in amplified regions. Hum. Genet. 56, 315–320.

Tardy, E.P., Tóth, A., 1997. Cross-hybridization of the chromosome 13/21 alpha satellite DNA to chromosome 22 or a rare polymorphism? Prenat. Diagn. 17, 487–488.

Tatton-Brown, K., Pilz, D.T., Orstavik, K.H., Patton, M., Barber, J.C., Collinson, M.N., Maloney, V.K., Huang, S., Crolla, J.A., Marks, K., Ormerod, E., Thompson, P., Nawaz, Z., Lese-Martin, C., Tomkins, S., Waits, P., Rahman, N., McEntagart, M., 2009. 15q overgrowth syndrome: A newly recognized phenotype associated with overgrowth, learning difficulties, characteristic facial appearance, renal anomalies and increased dosage of distal chromosome 15q. Am. J. Med. Genet. A 149A, 147–154.

Taysi, K., Chao, W.T., Monaghan, N., Monaco, M.P., 1983. Trisomy 6q22 leads to 6qter due to maternal 6;21 translocation. Case report review of the literature. Ann. Genet. 26, 243–246.

Tchatchou, S., Burwinkel, B., 2008. Chromosome copy number variation and breast cancer risk. Cytogenet. Genome Res. 123, 183–187.

Tepperberg, J., Pettenati, M.J., Rao, P.N., Lese, C.M., Rita, D., Wyandt, H., Gersen, S., White, B., Schoonmaker, M.M., 2001. Prenatal diagnosis using interphase fluorescence in situ hybridization (FISH): 2-year multi-center retrospective study and review of the literature. Prenat. Diagn. 21, 293–301.

Thilaganathan, B., Sairam, S., Ballard, T., Peterson, C., Meredith, R., 2000. Effectiveness of prenatal chromosomal analysis using multicolor fluorescent in situ hybridisation. BJOG 107, 262–266.

Tijo, J.H., Levan, A., 1956. The chromosome number in man. Hereditas 42, 1–6.

Tjio, J.H., Puck, T.T., Robinson, A., 1960. The human chromosomal satellites in normal persons and in two patients with Marfan's syndrome. Proc. Natl. Acad. Sci. USA 46, 532–539.

Till, M., Rafat, A., Charrin, C., Plauchu, H., Germain, D., 1991. Duplication of chromosome 11 centromere in fetal and maternal karyotypes: A new variant? Prenat. Diagn. 11, 481–482.

Thangavelu, M., Chen, P.X., Pergament, E., 1998. Hybridization of chromosome 18 alpha-satellite DNA probe to chromosome 22. Prenat. Diagn. 18, 922–925.

Thompson, P.W., Roberts, S.H., Rees, S.M., 1990. Replication studies in the 16p+ variant. Hum. Genet. 84, 371–372.

Tonk, V.S., Wilson, G.N., Yatsenko, S.A., Stankiewicz, P., Lupski, J.R., Schutt, R.C., Northup, J.K., Velagaleti, G.V., 2005. Molecular cytogenetic characterization of a familial der(1)del(1)(p36.33)dup(1)(p36.33p36.22) with variable phenotype. Am. J. Med. Genet. A 139A, 136–140.

Tonk, V., Kyhm, J.H., Gibson, C.E., Wilson, G.N., 2011. Interstitial deletion 5q14.3q21.3 with MEF2C haploinsufficiency and mild phenotype: When more is less. Am. J. Med. Genet. A 155A, 1437–1441.

Trabalza, N., Furbetta, M., Rosi, G., Donti, E., Venti, G., 1978. Migliorini Bruschelli G. Karyotype 46, XY,22p+ in a male patient. J. Genet. Hum. 26, 177–184.

Trifonov, V., Seidel, J., Starke, H., Martina, P., Beensen, V., Ziegler, M., Hartmann, I., Heller, A., Nietzel, A., Claussen, U., Liehr, T., 2003. Enlarged chromosome 13 p-arm hiding a cryptic partial trisomy 6p22.2-pter. Prenat. Diagn. 23, 427–430.

Trowell, H.E., Nagy, A., Vissel, B., Choo, K.H., 1993. Long-range analyses of the centromeric regions of human chromosomes 13, 14 and 21: Identification of a narrow domain containing two key centromeric DNA elements. Hum. Mol. Genet. 2, 1639–1649.

Tsita, K.P., Vallas, O.S., Velissariou, P.J., Lyberatou-Moraitou, E.K., 1989. A case of prenatal diagnosis of a familial satellited Yq chromosome. Clin. Genet. 35, 70–74.

Tsuchiya, K., Schueler, M.G., Dev, V.G., 2001. Familial X centromere variant resulting in false-positive prenatal diagnosis of monosomy X by interphase FISH. Prenat. Diagn. 21, 852–855.

Tüür, S., Käosaar, M., Mikelsaar, A.V., 1974. 1q plus variants in a normal adult population (one with a pericentric inversion). Humangenetik 24, 217–220.

Tyler-Smith, C., Brown, W.R., 1987. Structure of the major block of alphoid satellite DNA on the human Y chromosome. J. Mol. Biol. 195, 457–470.

Tyson, C., Harvard, C., Locker, R., Friedman, J.M., Langlois, S., Lewis, M.E., Van Allen, M., Somerville, M., Arbour, L., Clarke, L., McGilivray, B., Yong, S.L., Siegel-Bartel, J., Rajcan-Separovic, E., 2005. Submicroscopic deletions and duplications in individuals with intellectual disability detected by array-CGH. Am. J. Med. Genet. A 139, 173–185.

Udayakumar, A.M., Pathare, A.V., Dennison, D., Raeburn, J.A., 2009. Acquired pericentric inversion of chromosome 9 in acute myeloid leukemia. J. Appl. Genet. 50, 73–76.

Uehara, S., Akai, Y., Takeyama, Y., Takabayashi, T., Okamura, K., Yajima, A., 1992. Pericentric inversion of chromosome 9 in prenatal diagnosis and infertility. Tohoku. J. Exp. Med. 166, 417–427.

Vejerslev, L.O., Friedrich, U., 1984. Experiences with unexpected structural chromosome aberrations in prenatal diagnosis in a Danish series. Prenat. Diagn. 4, 181–186.

Vekemans, M., Morichon-Delvallez, N., 1990. Duplication of the long arm of chromosome 13 secondary to a recombination in a maternal intrachromosomal insertion (shift). Prenat. Diagn. 10, 787–794.

Venter, J.C., Adams, M.D., Myers, E.W., Li, P.W., Mural, R.J., Sutton, G.G., Smith, H.O., Yandell, M., Evans, C.A., Holt, R.A., Gocayne, J.D., Amanatides, P., Ballew, R.M., Huson, D.H., Wortman, J.R., Zhang, Q., Kodira, C.D., Zheng, X.H., Chen, L., Skupski, M., Subramanian, G., Thomas, P.D., Zhang, J., Gabor Miklos, G.L., Nelson, C., Broder, S., Clark, A.G., Nadeau, J., McKusick, V.A., Zinder, N., Levine, A.J., Roberts, R.J., Simon, M., Slayman, C., Hunkapiller, M., Bolanos, R., Delcher, A., Dew, I., Fasulo, D., Flanigan, M., Florea, L., Halpern, A., Hannenhalli, S., Kravitz, S., Levy, S., Mobarry, C., Reinert, K., Remington, K., Abu-Threideh, J., Beasley, E., Biddick, K., Bonazzi, V., Brandon, R., Cargill, M., Chandramouliswaran, I., Charlab, R., Chaturvedi, K., Deng, Z., Di Francesco, V., Dunn, P., Eilbeck, K., Evangelista, C., Gabrielian, A.E., Gan, W., Ge, W., Gong, F., Gu, Z., Guan, P., Heiman, T.J., Higgins, M.E., Ji, R.R., Ke, Z., Ketchum, K.A., Lai, Z., Lei, Y., Li, Z., Li, J., Liang, Y., Lin, X., Lu, F., Merkulov, G.V., Milshina, N., Moore, H.M., Naik, A.K., Narayan, V.A., Neelam, B., Nusskern, D., Rusch, D.B., Salzberg, S., Shao, W., Shue, B., Sun, J., Wang, Z., Wang, A., Wang, X., Wang, J., Wei, M., Wides, R., Xiao, C., Yan, C., Yao, A., Ye, J., Zhan, M., Zhang, W., Zhang, H., Zhao, Q., Zheng, L., Zhong, F., Zhong, W., Zhu, S., Zhao, S., Gilbert, D., Baumhueter, S., Spier, G., Carter, C., Cravchik, A., Woodage, T., Ali, F., An, H., Awe, A., Baldwin, D., Baden, H., Barnstead, M., Barrow, I., Beeson, K., Busam, D., Carver, A., Center, A., Cheng, M.L., Curry, L., Danaher, S., Davenport, L., Desilets, R., Dietz, S., Dodson, K., Doup, L., Ferriera, S., Garg, N., Gluecksmann, A., Hart, B., Haynes, J., Haynes, C., Heiner, C., Hladun, S., Hostin, D., Houck, J., Howland, T., Ibegwam, C., Johnson, J., Kalush, F., Kline, L., Koduru, S., Love, A., Mann, F., May, D., McCawley, S., McIntosh, T., McMullen, I., Moy, M., Moy, L., Murphy, B., Nelson, K., Pfannkoch, C., Pratts, E., Puri, V., Qureshi, H., Reardon, M., Rodriguez, R., Rogers, Y.H., Romblad, D., Ruhfel, B., Scott, R., Sitter, C., Smallwood, M., Stewart, E., Strong, R., Suh, E., Thomas, R., Tint, N.N., Tse, S., Vech, C., Wang, G., Wetter, J., Williams, S., Williams, M., Windsor, S., Winn-Deen, E., Wolfe, K., Zaveri, J., Zaveri, K., Abril, J.F., Guigó, R., Campbell, M.J., Sjolander, K.V., Karlak, B., Kejariwal, A., Mi, H., Lazareva, B., Hatton, T., Narechania, A., Diemer, K., Muruganujan, A., Guo, N., Sato, S., Bafna, V., Istrail, S., Lippert, R., Schwartz, R., Walenz, B., Yooseph, S., Allen, D., Basu, A., Baxendale, J., Blick, L., Caminha, M., Carnes-Stine, J., Caulk, P., Chiang, Y.H., Coyne, M., Dahlke, C., Mays, A., Dombroski, M., Donnelly, M., Ely, D., Esparham, S., Fosler, C., Gire, H., Glanowski, S., Glasser, K., Glodek, A., Gorokhov, M., Graham, K.,

Gropman, B., Harris, M., Heil, J., Henderson, S., Hoover, J., Jennings, D., Jordan, C., Jordan, J., Kasha, J., Kagan, L., Kraft, C., Levitsky, A., Lewis, M., Liu, X., Lopez, J., Ma, D., Majoros, W., McDaniel, J., Murphy, S., Newman, M., Nguyen, T., Nguyen, N., Nodell, M., Pan, S., Peck, J., Peterson, M., Rowe, W., Sanders, R., Scott, J., Simpson, M., Smith, T., Sprague, A., Stockwell, T., Turner, R., Venter, E., Wang, M., Wen, M., Wu, D., Wu, M., Xia, A., Zandieh, A., Zhu, X., 2001. The sequence of the human genome. Science 291, 1304–1351.

Velázquez, M., Visedo, G., Ludeña, P., de Cabo, S.F., Sentís, C., Fernández-Piqueras, J., 1991. Cytogenetic analysis of a human familial 15p+ marker chromosome. Genome 34, 827–829.

Verellen-Dumoulin, C., Freud, M., De Meyer, R., Laterre, C., Frederic, J., Thompson, M.W., Markovic, V.D., 1984. Expression of an X-linked muscular dystrophy in a female due to translocation involving Xp21 and non-random inactivation of the normal X chromosome. Hum. Genet. 67, 115–119.

Verlinsky, Y., Ginsberg, N., Chmura, M., Freidine, M., White, M., Strom, C., Kuliev, A., 1995. Cross-hybridization of the chromosome 13/21 alpha satellite DNA probe to chromosome 22 in the prenatal screening of common chromosomal aneuploidies by FISH. Prenat. Diagn. 15, 831–834.

Verma, R.S., Babu, A., 1987. 18ph+ is a so-called normal chromosomal variant. Clin. Genet. 32, 419–420.

Verma, R.S., Dosik, H., Lubs, H.A., 1977. Size variation polymorphisms of the short arm of human acrocentric chrosomes determined by R-banding by fluorescence using acridine orange (RFA). Hum. Genet. 38, 231–234.

Verma, R.S., Dosik, H., Jhaveri, R.C., Warman, J., 1978. Cytogenetic polymorphism or Y/15 translocation in a black male with ambiguous genitalia. J. Genet. Hum. 26, 405–409.

Verma, R.S., Dosik, H., Lubs, H.A., 1978. Size and pericentric inversion heteromorphisms of secondary constriction regions (h) of chromosomes 1, 9, and 16 as detected by CBG technique in Caucasians: classification, frequencies, and incidence. Am. J. Med. Genet. 2, 331–339.

Verma, R.S., Ved Brat, S., Warman, J., Dosik, H., 1979. Clinical significance of the satellited short arm of human chromosome 17 (17ps +): A rare heteromorphism? Ann. Genet. 22, 133–136.

Verma, R.S., Rodriguez, J., Dosik, H., 1981. Human chromosome heteromorphisms in Americans Blacks: II. Higher incidence of pericentric inversions of secondary constriction regions (h). Am. J. Med. Genet. 8, 17–25.

Verma, R.S., Huq, A., Dosik, H., 1983. Racial variation of a nonfluorescent segment of the Y chromosome in East Indians. J. Med. Genet. 20, 102–106.

Verma, R.S., Luke, S., Conte, R.A., Macera, M.J., 1991. A so-called rare heteromorphism of the human genome. Cytogenet. Cell Genet. 56, 63.

Verma, R.S., Luke, S., Mathews, T., Conte, R.A., 1992. Molecular characterization of the smallest secondary constriction region (qh) of human chromosome 16. Genet. Anal. Tech. Appl. 9, 140–142.

Verma, R.S., Luke, S., Brennan, J.P., Mathews, T., Conte, R.A., Macera, M.J., 1993. Molecular topography of the secondary constriction region (qh) of human chromosome 9 with an unusual euchromatic band. Am. J. Hum. Genet. 52, 981–986.

Verma, R.S., Kleyman, S.M., Conte, R.A., 1996. Molecular characterization of an unusual variant of the short arm of chromosome 15 by FISH-technique. Jpn. J. Hum. Genet. 41, 307–311.

Verma, R.S., Ramesh, K.H., Mathews, T., Kleyman, S.M., Conte, R.A., 1996. Centromeric alphoid sequences are breakage prone resulting in pericentromeric inversion heteromorphism of qh region of chromosome 1. Ann. Genet. 39, 205–208.

Verma, R.S., Gogineni, S.K., Kleyman, S.M., Conte, R.A., 1997. Characterisation of a satellited non-fluorescent Y chromosome (Y[nfqs]) by FISH. J. Med. Genet. 34, 817–818.

Verma, R.S., Batish, S.D., Gogineni, S.K., Kleyman, S.M., Stetka, D.G., 1997. Centromeric alphoid DNA heteromorphisms of chromosome 21 revealed by FISH-technique. Clin. Genet. 51, 91–93.

Verma, R.S., Ishwar, L., Gogineni, S.K., Kleyman, S.M., 1998. Pericentromeric heteromorphism of human chromosome 18 as revealed by FISH-technique. Ann. Genet. 41, 154–156.

Vermeesch, J.R., Duhamel, H., Raeymaekers, P., Van Zand, K., Verhasselt, P., Fryns, J.P., Marynen, P., 2003. A physical map of the chromosome 12 centromere. Cytogenet. Genome Res. 103, 63–73.

Verschuuren-Bemelmans, C.C., Leegte, B., Hodenius, T.M., Cobben, J.M., 1995. Trisomy 1q42 –> qter in a sister and brother: further delineation of the "trisomy 1q42 –> qter syndrome". Am. J. Med. Genet. 58, 83–86.

Vialard, F., Molina-Gomes, D., Roume, J., Podbiol, A., Hirel, C., Bailly, M., Hammoud, I., Dupont, J.M., de Mazancourt, P., Selva, J., 2009. Case report: Meiotic segregation in spermatozoa of a 46, X, t(Y;10)(q11.2;p15.2) fertile translocation carrier. Reprod. Biomed. Online 18, 549–554.

Villa, N., Sala, E., Colombo, D., Dell'Orto, M., Dalprà, L., 2000. Monosomy and trisomy 1q44-qter in two sisters originating from a half cryptic 1q;15p translocation. J. Med. Genet. 37, 612–615.

Vorsanova, S.G., Yurov, Y.B., Brusquant, D., Carles, E., Roizes, G., 2002. Two new cases of the Christchurch (Ch1) chromosome 21: Evidence for clinical consequences of de novo deletion 21P-. Tsitol. Genet. 36, 46–49.

Vust, A., Riordan, D., Wickstrom, D., Chudley, A.E., Dawson, A.J., 1998. Functional mosaic trisomy of 1q12–>1q21 resulting from X-autosome insertion translocation with random inactivation. Clin. Genet. 54, 70–73.

Wachtler, F., Musil, R., 1980. On the structure and polymorphism of the human chromosome no. 15. Hum. Genet. 56, 115–118.

Waldeyer, W., 1888. Über Karyogenese und ihre Beziehung zu den Befruchtungsvorgängen. Arch. Mikrosc. Anat. 32, 1–22.

Wahedi, K., Pawlowitzki, I.H., 1987. C-band polymorphisms of chromosome 9: Quantification by Ce-bands. Hum. Genet. 77, 1–5.

Walker, L.C., Krause, L., 2012. kConFab Investigators, Spurdle AB, Waddell N. Germline copy number variants are not associated with globally acquired copy number changes in familial breast tumours. Breast Cancer Res. Treat. 134, 1005–1011.

Wall, W.J., Clark, M.S., Coates, P., 1988. Structural complexity of Y chromosome heterochromatin. Cytobios. 56, 17–22.

Walzer, S., Breau, G., Gerald, P.S., 1969. A chromosome survey of 2,400 normal newborn infants. J. Pediatr. 74, 438–448.

Wan, T.S., Ma, S.K., Chan, L.C., 2000. Acquired pericentric inversion of chromosome 9 in essential thrombocythemia. Hum. Genet. 106, 669–670.

Wang, J.C., Fisker, T., Dang, L., Teshima, I., Nowaczyk, M.J., 2009. 4.3-Mb triplication of 4q32.1-q32.2: report of a family through two generations. Am. J. Med. Genet. A 149A, 2274–2279.

Warburton, P.E., Hasson, D., Guillem, F., Lescale, C., Jin, X., Abrusan, G., 2008. Analysis of the largest tandemly repeated DNA families in the human genome. BMC Genomics 9, 533.

Watson, E.J., Scrimgeour, J.B., 1977. A dicentric no. 15 chromosome with two satellite regions. J. Med. Genet. 14, 381–383.

Watt, J.L., Couzin, D.A., Lloyd, D.J., Stephen, G.S., McKay, E., 1984. A familial insertion involving an active nucleolar organiser within chromosome 12. J. Med. Genet. 21, 379–384.

Waye, J.S., Greig, G.M., Willard, H.F., 1987. Detection of novel centromeric polymorphisms associated with alpha satellite DNA from human chromosome 11. Hum. Genet. 77, 151–156.

Waye, J.S., Willard, H.F., 1989. Chromosome specificity of satellite DNAs: Short- and long-range organization of a diverged dimeric subset of human alpha satellite from chromosome 3. Chromosoma. 97, 475–480.

Wei, S., Siu, V.M., Decker, A., Quigg, M.H., Roberson, J., Xu, J., Adeyinka, A., 2007. False-positive prenatal diagnosis of trisomy 18 by interphase FISH: Hybridization of chromosome 18 alpha-satellite DNA probe (D18Z1) to the heterochromatic region of chromosome 9. Prenat. Diagn. 27, 1064–1066.

Weier, H.U., Gray, J.W., 1992. A degenerate alpha satellite probe, detecting a centromeric deletion on chromosome 21 in an apparently normal human male, shows limitations of the use of satellite DNA probes for interphase ploidy analysis. Anal. Cell Pathol. 4, 81–86.

Weise, A., Starke, H., Mrasek, K., Claussen, U., Liehr, T., 2005. New insights into the evolution of chromosome 1. Cytogenet. Genome Res. 108, 217–222.

Weise, A., Gross, M., Mrasek, K., Mkrtchyan, H., Horsthemke, B., Jonsrud, C., Von Eggeling, F., Hinreiner, S., Witthuhn, V., Claussen, U., Liehr, T., 2008. Parental-origin-determination fluorescence in situ hybridization distinguishes homologous human chromosomes on a single-cell level. Int. J. Mol. Med. 21, 189–200.

Weise, A., Mrasek, K., Klein, E., Mulatinho, M., Llerena Jr., J.C., Hardekopf, D., Pekova, S., Bhatt, S., Kosyakova, N., Liehr, T., 2012. Microdeletion and microduplication syndromes. J. Histochem. Cytochem. 60, 346–358.

Weiss, K.M., 1998. In search of human variation. Genome Res. 8, 691–697.

Wellauer PK, Dawid IB. Isolation and sequence organization of human ribosomal DNA. J. Mol. Biol. 197; 128:289-303.

Wevrick, R., Willard, H.F., 1991. Physical map of the centromeric region of human chromosome 7: Relationship between two distinct alpha satellite arrays. Nucleic Acids Res. 19, 2295–2301.

Weremowicz, S., Sandstrom, D.J., Morton, C.C., Niedzwiecki, C.A., Sandstrom, M.M., Bieber, F.R., 2001. Fluorescence in situ hybridization (FISH) for rapid detection of aneuploidy: Experience in 911 prenatal cases. Prenat. Diagn. 21, 262–269.

Werner, W., Herrmann, F.H., 1984. Analysis of a familial 15p + polymorphism: Exclusion of Y/15 translocation. Clin. Genet. 26, 204–208.

Whiteford, M.L., Baird, C., Kinmond, S., Donaldson, B., Davidson, H.R., 2000. A child with bisatellited, dicentric chromosome 15 arising from a maternal paracentric inversion of chromosome 15q. J. Med. Genet. 37, E11.

Wilkinson, T.A., Crolla, J.A., 1993. Molecular cytogenetic characterization of three familial cases of satellited Y chromosomes. Hum. Genet. 91, 389–391.

Willard, H.F., 1991. Evolution of alpha satellite. Curr. Opin. Genet. Dev. 1, 509–514.

Willard, F.H., Waye, J.S., 1987. Hierarchical order in chromosome-specific human alpha satellite DNA. Trends Genet. 3, 192–198.

Willard, H.F., Greig, G.M., Powers, V.E., Waye, J.S., 1987. Molecular organization and haplotype analysis of centromeric DNA from human chromosome 17: Implications for linkage in neurofibromatosis. Genomics 1, 368–373.

Willatt, L., Green, A.J., Trump, D., 2001. Satellites on the terminal short arm of chromosome 12 (12ps), inherited through several generations in three families: A new variant without phenotypic effect. J. Med. Genet. 38, 723–727.

Wilkie, A.O., Higgs, D.R., Rack, K.A., Buckle, V.J., Spurr, N.K., Fischel-Ghodsian, N., Ceccherini, I., Brown, W.R., Harris, P.C., 1991. Stable length polymorphism of up to 260 kb at the tip of the short arm of human chromosome 16. Cell 64, 595–606.

Winsor, E.J., Dyack, S., Wood-Burgess, E.M., Ryan, G., 1999. Risk of false-positive prenatal diagnosis using interphase FISH testing: Hybridization of alpha-satellite X probe to chromosome 19. Prenat. Diagn. 19, 832–836.

Yamada, K., 1992. Population studies of inv(9) chromosomes in 4,300 Japanese: Incidence, sex difference and clinical significance. Jpn. J. Hum. Genet. 37, 293–301.

Yang, J., Schwinger, E., Mennicke, K., 2001. Primed in situ labeling: Sensitivity and specificity for detection of alpha-satellite DNA in the centromere regions of chromosomes 13 and 21. Cytogenet. Cell Genet. 95, 28–33.

Yoshida, A., Nakahori, Y., Kuroki, Y., Miura, K., Shirai, M., 1997. An azoospermic male with an unbalanced autosomal-Y translocation. Jpn. J. Hum. Genet. 42, 451–455.

Yurov, Y.B., Mitkevich, S.P., Alexandrov, I.A., 1987. Application of cloned satellite DNA sequences to molecular-cytogenetic analysis of constitutive heterochromatin heteromorphisms in man. Hum. Genet. 76, 157–164.

Zankl, H., Zang, K.D., 1971. Structural variability of the normal human karyotype. Humangenetik 13, 160–162.

Zaslav, A.L., Pierno, G., Fougner, A., Jacob, J., Shikora, G., Kazi, R., Blumenthal, D., Alexander, F., Fox, J.E., 2004. Prenatal diagnosis of a rare inherited heterochromatic variant chromosome 4. Am. J. Med. Genet. A 126A, 420–422.

Zelante, L., Notarangelo, A., Dallapiccola, B., 1994. The 18ph+ chromosome heteromorphism. Prenat. Diagn. 14, 1096–1097.

Zhao, J., Gordon, P.L., Wilroy Jr., R.S., Martens, P.R., Tarleton, J., Shulman, L.P., Simpson, J.L., Elias, S., Tharapel, A.T., 1995. Characterization of an unbalanced de novo rearrangement by microsatellite polymorphism typing and by fluorescent in situ hybridization. Am. J. Med. Genet. 56, 398–402.

Zhao, L., Li, H., Xue, Y.Q., Pan, J.L., Wu, Y.F., Lu, M., 2004. Application of fluorescence in situ hybridization in the diagnosis of genetic diseases. Zhonghua Yi Xue Yi Chuan Xue Za Zhi 21, 611–614.

Zhuang, J., Hu, X., Zhang, B., 1994. Chromosome 15 satellite enlargement and hereditary deafness. Zhonghua Er Bi Yan Hou Ke Za Zhi 29, 225–227.

INDEX

Note: Page numbers with "*f*" denote figures; "*t*" tables; and "*b*" boxes.

A

aCGH. *See* array-based comparative genomic hybridization

Acrocentric chromosomes' short arms, 37

addition
  to nonacrocentric p-arms, 48
  parental cytogenetic analysis, 49b
  to q-arms, 48
  reports on derivatives, 49
  translocation, 48–49
altered acrocentric p-arms, 52
  p+ variants, 52
  reported addition of euchromatin, 53t
Band p11.2, 38
classical satellite DNAs, 38–39
components, 38
enlargement, 39–40
heterochromatic material addition, 44–48
insertion, 49–50
  effect on gene, 50
  inversion loop formation, 50f
  parental cytogenetic analysis, 50b
length of short arms with, 14t
loss of material, 51
  NOR or CBG staining, 52b
  p– variants, 51–52
material amplification, 40
  acrocentric p-arm material amplification, 41t
  acrocentric p13 material amplification, 40–43
  CG-CNV stkstk/ss, 43
  nomenclature for heterochromatic variants, 43
  NOR or CBG staining use, 44b
  p+ variants, 43–44
  short arm variants as idiogram, 42f

normal size of, 40f
positions and sizes in, 15t
satellite I DNA, 38
schematic depictions of, 39f
suggestion for normal/average length, 13

Acrocentric chromosomes' short-arm variants, 29, 30f
acrocentric short arms, 13
  length of short arms with, 14t
  positions and Sizes in, 15t
amplification or loss of short-arm material, 30
genome browsers, 13
ISCN, 13
mitotic translocation event, 30
NOR, 13
translocations, 29
Yq12 heterochromatin translocation, 29

α-satellite DNA, 38–39
CG-CNV, 73
D9Z4, 65–66
DNA stretch DYZ3, 80
FISH analysis, 56
size variants, 58
of Y chromosome, 57

Alphoid probe
D13Z1, 70
D14Z1, 71
D18Z1, 75
D8Z2, 65
p4n1/4 and pG-Xba11/340, 62, 65–66
2cen+, 60

Angelman syndrome (AS), 98

array-based comparative genomic hybridization (aCGH), 1, 124
MG-CNVs in, 4f, 126
molecular genetics using, 122
region harbors segmental duplications, 93–94

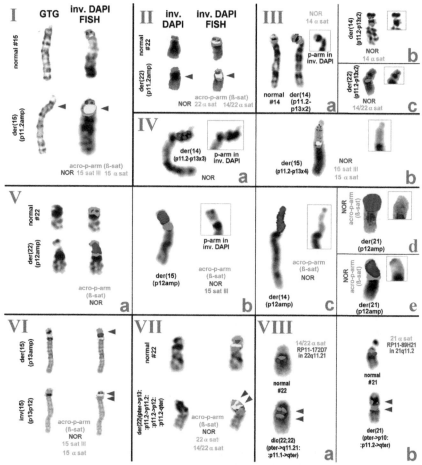

Color Plate 1 Variants of the acrocentric short arms; arrowheads indicate the aberrations and/or breakpoints. Abbreviations: DAPI = 4,6-diamino-2phenyl-indole/distamycin A staining/banding; GTG = Giemsa-banding; inv. = inverted I. Normal and derivative chromosome 15 with enlarged short arm due to D15Z1 amplification (yellow signal). II. Normal and derivative chromosome 22 with enlarged short arm due to D22Z4 amplification (yellow signal). III. Normal and derivative chromosome 14 with enlarged short arm due to two NOR regions (a), which look in inverted DAPI similar to enlarged p-arms of a chromosome 22 (c), but different to another chromosome 14p-arm (b). IV. Enlargement of the short arm of chromosome 14 due to NOR triplication (a), and of a chromosome 15 by a NOR quadruplication (b). V. Enlargement of the short arms and of chromosome 22 (a), 15 (b), 14 (c), and 21 (d, e) due to NOR amplification. VI. In one person an inv p13 enlarged chromosome 15 and an inversion within the other short arm could be observed. VII. Enlargement of the short arm of chromosome 22 due to a complex inverted duplication. VIII. Examples for (pseudo-)dicentric chromosomes 22 (a) and 21 (b) leading to enlarged short arm in comparison to normal homologous chromosomes.

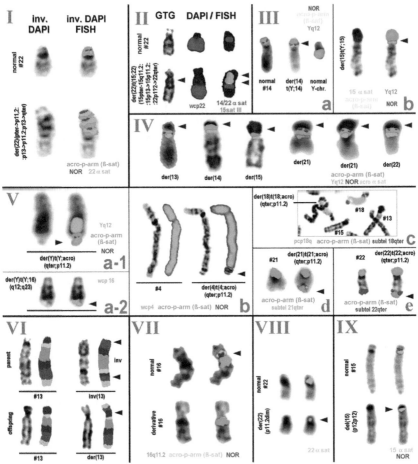

Color Plate 2 Variants of the acrocentric short arms; arrowheads indicate the aberrations and/or breakpoints. Abbreviations: DAPI = 4,6-diamino-2phenyl-indole/distamycin A staining/banding; GTG = Giemsa-banding; inv. = inverted I. Normal and derivative chromosome 22 with massively enlarged short arm due to partial triplication. II. Normal and derivative chromosome 22 with massively enlarged short arm due to translocation and duplication of a part of chromosome 15. III. Altered/enhanced short arms of chromosomes 14 (a) and 15 (b) in comparison to a normal homologous chromosome and a Y chromosome (a). Yq12-material was translocated in both cases to the derivative chromosomes. IV. Altered/enhanced short arms of chromosomes 13, 14, 15, 21, and 22 of different lengths due to addition of amplified material of unknown origin. V. Addition of acrocentric short arm material to the ends of other chromosomes like the Y chromosome (a-1), chromosome 4 (b), 18 (c), 21 (d), or 22 (e). a-2 depicts a derivative Y chromosome with parts of chromosome 16 attached to the band Yq12; this leads to a partial trisomy 16q23 to 16qter and is no CG-CNV. However, in banding cytogenetics such a derivative Y chromosome may be mixed up with a heteromorphic variant as in a-1. VI. Inverted DAPI banding shown together with multicolor banding (MCB) results [Liehr et al., 2010] of a parent and its offspring having a normal chromosome 15 and a derivative one, each. For further explanations see Figure 15. VII. Insertion of acrocentric short arm material into chromosome 16p. VIII. Diminishing of an acrocentric short arm due to loss of (parts of) band p11.2, exemplified on chromosome 22. IX. Diminishing of an acrocentric short arm due to loss of (parts of) band p12, exemplified on chromosome 15.

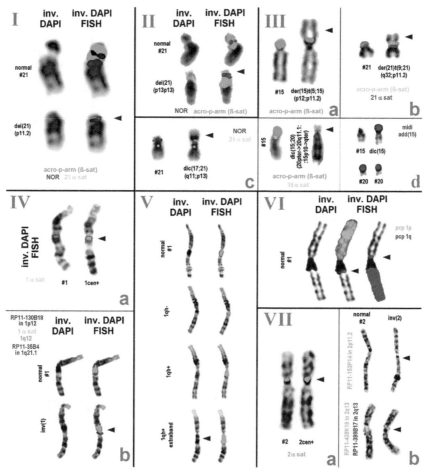

Color Plate 3 Variants of the acrocentric short arms (I–III) and the centromeric regions (IV–VII); arrowheads indicate the aberrations and/or breakpoints. Abbreviations: DAPI = 4,6-diamino-2phenyl-indole/distamycin A staining/banding; GTG = Giemsa-banding; inv. = inverted; midi = microdissection and reverse FISH I. Complete loss of an acrocentric short arm exemplified on chromosome 21 and compared to a normal one. II. Diminishing of an acrocentric short arm due to loss of (parts of) band p13, exemplified on chromosome 21. III. Enlargements of acrocentric short arms with clinical impact; parts of chromosome 5 (a), 9 (b), 17 (c), or 20 (d) were added. IV. Reduction of alphoid DNA stretches in comparison to a normal-sized stretch (a) and pericentric inversion polymorphism (b) of chromosome 1. V. Normal size, reduction, enlargement, and huge enlargement of the band 1q12. VI. There is a cross-hybridization of partial chromosome paint (pcp) 1q in centromere-near 1p and of pcp 1p in centromere-near 1q due to homologous sequences. VII. Two typical heteromorphisms of chromosome 2: (a) a stronger than normal signal of the alphoid probe leading to a 2ph+ in banding cytogenetics and (b) a pericentric inversion polymorphism.

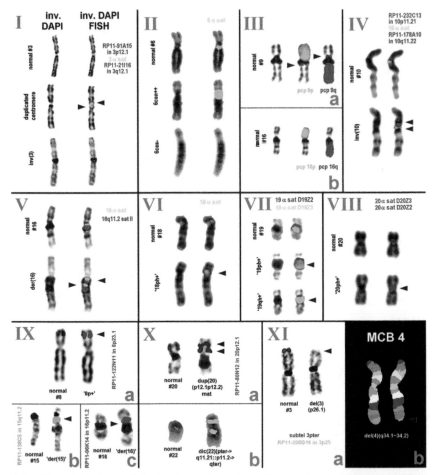

Color Plate 4 Variants of the centromeric regions (I-); arrowheads indicate the aberrations and/or breakpoints. Abbreviations: DAPI = 4,6-diamino-2phenyl-indole/distamycin A staining/banding; GTG = Giemsa-banding; inv. = inverted I. A duplicated centromere and an inversion heteromorphism of chromosome 3 compared to a normal one. II. Extreme enlargement of a centromere of a chromosome 6 and a 6cen- compared to a normal one. III. Similar FISH-experiment as depicted in Color plate 3, Figure V, for chromosomes 9 (a) and 16 (b). There is a cross-hybridization of partial chromosome paint (pcp) 9q in centromere-near 9p and of pcp 9p in centromere-near 9q due to homologous sequences (arrowheads); no such cross-hybridization is visible for chromosome 16 applying the corresponding pcp probes. IV. Nondeleterious inversion inv(10)(p11.2q21.2) observed in chromosome 10 and highlighted using centromeric and locus-specific centromere-near BAC probes. V. Normal and derivative chromosomes 16 are depicted; in inverted DAPI banding an extraband is visible in the subcentromeric region of the long arm. Using appropriate probes for 16p11.1-q11.1 (green) and a satellite II probe for 16p11.2, a pseudodicentric chromosome der(16)(pter->q11.2::p11.1->qter) could be defined. VI. Cen+ variant of chromosome 18 together with a normal-sized D18Z1 stained centromere of chromosome 18. The enlargement of the D18Z1-stretch went into the short arm of the corresponding chromosome 18; thus, these derivatives are described in the literature as well as 18ph+. VII. A normal chromosome 19 together with two heteromorphic chromosomes 19 from two different patients is depicted. In both cases the satellite stretch D19Z3 is enlarged but not D19Z2. However, one heteromorphic chromosome 19 appears to be enlarged in the short arm (ph+) and the other in the long arm (qh+), according to inverted DAPI-banding. VIII. In heteromorphic chromosomes 20 only enlargement of D20Z3 and not D20Z2 was observed. Sometimes derivative chromosomes 20 as depicted here are referred as 20ph+. IX. Euchromatic variants (EVs) in 8p23.1 (a), 15q11.2 (b), and 16p11.2 (c) are proven by FISH using the appropriate probes. X. Euchromatic UBCAs expressed as duplications without clinical impact for their carriers: (a) a large duplication in the short arm of chromosome 20 and (b) a pseudodicentric chromosome 22 broken in 22p11.2 and 22q11.21. XI. Euchromatic UBCAs expressed as terminal deletions without clinical impact for their carriers: (a) a deletion in the short arm of chromosome 3 highlighted by locus-specific BAC-probes and (b) a deletion in the long arm of chromosome 4, depicted in multicolor-banding (MCB) pseudo-coloring pattern.